Studies in Systems, Decision and Control

Volume 407

The series "Studies in Systems, Decision and Control" (SSDC) covers both new developments and advances, as well as the state of the art, in the various areas of broadly perceived systems, decision making and control–quickly, up to date and with a high quality. The intent is to cover the theory, applications, and perspectives on the state of the art and future developments relevant to systems, decision making, control, complex processes and related areas, as embedded in the fields of engineering, computer science, physics, economics, social and life sciences, as well as the paradigms and methodologies behind them. The series contains monographs, textbooks, lecture notes and edited volumes in systems, decision making and control spanning the areas of Cyber-Physical Systems, Autonomous Systems, Sensor Networks, Control Systems, Energy Systems, Automotive Systems, Biological Systems, Vehicular Networking and Connected Vehicles, Aerospace Systems, Automation, Manufacturing, Smart Grids, Nonlinear Systems, Power Systems, Robotics, Social Systems, Economic Systems and other. Of particular value to both the contributors and the readership are the short publication timeframe and the world-wide distribution and exposure which enable both a wide and rapid dissemination of research output.

Indexed by SCOPUS, DBLP, WTI Frankfurt eG, zbMATH, SCImago.

All books published in the series are submitted for consideration in Web of Science.

More information about this series at https://link.springer.com/bookseries/13304

Anand J. Kulkarni
Editor

Multiple Criteria Decision Making

Techniques, Analysis and Applications

Springer

Editor
Anand J. Kulkarni
Institute of Artificial Intelligence
MIT World Peace University
Pune, India

ISSN 2198-4182 ISSN 2198-4190 (electronic)
Studies in Systems, Decision and Control
ISBN 978-981-16-7413-6 ISBN 978-981-16-7414-3 (eBook)
https://doi.org/10.1007/978-981-16-7414-3

This Springer imprint is published by the registered company Springer Nature Singapore Pte Ltd.
The registered company address is: 152 Beach Road, #21-01/04 Gateway East, Singapore 189721, Singapore

Anand would like to dedicate this book to the memories of his good friend Moumen Darcherif, former Vice President of the Paris-Seine University, France. The book was conceptualized with Moumen; however, sadly lost him during the journey.

Preface

Multiple criteria decision-making (MCDM) provides a mathematical methodology that incorporates the values of decision-makers and stakeholders along with the technical information to select the best solution for a given problem. In general, decision-making occurs in all domains; however, it is challenging to deal with different tangible and intangible alternatives from diverse aspects with an intention of selecting the best one amongst them. The MCDM techniques are indispensable for decision-makers as the evaluation of a problem with different alternatives has conflicting and incommensurable objectives and is required to be optimized simultaneously. Importantly, the use of these techniques reduces subjectivity because of the psychology–human interaction. The aim of MCDM is to obtain the optimal alternative/choice that has the highest degree of satisfaction for most of the criteria. The growing recognition of MCDM approaches has motivated several researchers to further develop, test and apply them in various fields. In recent times, digital automation, machine learning, big data, IoT and artificial-intelligence-based methods offer promising solutions to the growing complexity. Today's strategic decision-making needs to be re-evaluated and addressed through advanced MCDM and integrated approaches of AI, big data and IoT, to provide more realistic and robust solutions to the current problems.

The edited book intends to provide a platform for interdisciplinary state-of-the-art discussion on MCDM with a focus on critical literature, underlying principles of methods and models, solution approaches, testing and validation, real-world applications, case studies etc. The book may provide guidelines to the potential MCDM researchers about the choice of approaches for dealing with the complexities and modalities. The contributions of the book may further help to explore new avenues leading towards multi-disciplinary research discussions.

Every chapter submitted to the book has been critically evaluated by at least two expert reviewers. The critical suggestions by the reviewers helped and influenced the authors of the individual chapter to enrich the quality in terms of experimentation, performance evaluation, representation etc. The book may serve as a valuable reference for the researchers working in the domain of MCDM.

Chapter "MIVES: A Multi-Attribute Value Function-Based Methodology for Sustainability Assessment" by Biswal et al. highlights that any process or product design depends on the triple bottom approach of economic, environment and social criteria; however, sustainability assessment remains an important aspect of it. It further emphasizes the necessity of handling the associated subjectivity and difficulty in assessment. Accordingly, the chapter discusses MIVES (Modelo Integrado de Valor para una Evaluación Sostenible), a multi-attribute value function-based methodology framework for the necessary sustainability assessments. The applications of this methodology in construction, aviation, education and biomass processing industries are thoroughly discussed, which further underscores the need for a framework for sustainability evaluation in different domains of various industries.

One of the very recent MCDM methods referred to as the base-criterion method (BCM) is thoroughly discussed by Haseli and Sheikh in chapter "Base Criterion Method (BCM)". The discussion includes the detailed theoretical and mathematical formulation of the BCM. The essential characteristic of this method is that it removes a large number of unnecessary comparisons by dividing pairwise comparisons into two categories, viz., base comparisons and final comparisons. Importantly, the chapter provides illustrative examples showcasing the problem-solving process using the BCM method. The examples involving several criteria are associated with staff selection, product development and mode of transportation of products to the market. The examples are utilized to highlight the characteristics of the method; however, the chapter distinctively provides the limitations of the method along with recommendations.

DEX (Decision EXpert) is a method combining multi-criteria decision analysis and artificial intelligence suited particularly for sorting/classification decision problems. It is characterized by its hierarchical, qualitative, rule-based, multi-criteria decision modeling approaches. Chapter "DEX (Decision EXpert): A Qualitative Hierarchical Multi-criteria Method" by Marko Bohanec described DEX from the theoretical and practical viewpoint, and further explained it in terms of motivation, history, software, applications and method extensions. The work is supported and illustrated using three examples: a didactic example of employee selection and two real-world industrial applications of choosing a raw-material location and assessing electric energy production technologies, respectively. Importantly, the chapter in detail lists the limitations and the associated solutions as well.

The MCDM methods, such as the analytic hierarchy process (AHP) and the technique for order of preference by similarity to ideal solution (TOPSIS), work on the crisp data. The data may be inadequate and vague when dealing with real-world problems. This issue is addressed in chapter "Analysis of Fuzzy AHP and Fuzzy TOPSIS Methods for the Prioritization of the Software Requirements" by Nazim et al. by successfully applying them under a fuzzy environment in different areas, like management science, software engineering etc. The chapter attempts to present a comparison of the accuracy of the fuzzy AHP and fuzzy TOPSIS methods using different cases of software requirements prioritization problems. The chapter describes detailed mathematical formulations of the fuzzy AHP and fuzzy TOPSIS methods. The authors have also thrown light on limitations as well as possible enhancement of the study in terms of the number and size of the test cases.

Digital image forensic science is one of the very niche areas working towards checking the authenticity of digital images. Even though several algorithms have been developed to verify the forged images, the evaluation and selection of the digital image forensic tools based on different features, like error-level analysis, metadata analysis etc., are still required to be explored. This issue is addressed by Parveen et al. in chapter "A Fuzzy-Based Multi-Criteria Decision-Making Approach for the Selection of Digital Image Forensic Tools". It discusses an algorithm specifically developed for the selection of the digital image forensic tools based on the ranking values. The ranking values of the digital image forensic tools are computed using TOPSIS method by using the triangular fuzzy numbers. The utilization of the proposed method is discussed with the help of an example in which the tools, such as FotoForensics, JPEGsnoop, Forensically, Ghiro and Izitru, are utilized during the analysis.

The decision matrix in the multi-criteria decision analysis (MCDA) problem necessarily involves a set of criteria and alternatives. They may have different units of measurement and are not suitable for a direct comparison. In chapter "Why Does the Choice of Normalization Technique Matter in Decision-Making", Shekhovtsov et al. underscore this issue and highlight that most of the MCDA methods require normalization of the decision matrix which can contribute to change in the final result. The chapter provides a significant investigation to show the fundamental differences between the five most common and prominent normalization techniques, viz., the minimum–maximum method, the maximum method, the sum method, the vector method and the logarithmic method. The methods have been statistically compared by using six randomly generated diverse data sets, differing in range, size and sign. The chapter highlights that the characteristics of data sets have a significant impact on normalization results. It further highlights the need for further investigation in interval normalization as well as fuzzy numbers.

Aggregation–disaggregation or ordinal regression approach is currently considered as an important tool at the disposal of potential analysts and decision-makers when addressing MCDM problems. Chapter "Bipolar Multicriteria Aggregation-Disaggregation Robustness Approach: Theory and Application on European e-Government Benchmarking" by Siskos et al. proposed a bipolar robustness control approach, implemented in conjunction with the UTASTAR method characterized by an additive value function referred to as the multi-criteria evaluation model. The methodology is applied to the problem of an e-government readiness evaluation in Europe, resulting in the ranking of 22 European countries. This application is based on one of the author's earlier developed multi-criteria e-government modeling approaches. The authors bring attention to certain robustness issues of the ordinal regression framework and the way that the preferential parameters are estimated through the UTA-type inference engine as well. The chapter in detail discusses the additive value model and UTASTAR method, and the principle of bipolar ordinal regression process and the robustness control.

Chapter "The COMET Method: Study Case of Swimming Training Progress" by Więckowski and Watróbski in detail describes an MCDA technique referred to as characteristic objects method (COMET). The theory of fuzzy numbers combined with the COMET method is used to create an evaluation model with complete knowledge and a certain degree of inherent uncertainty. The COMET and linear regression method have been used in a practical application in determining the trend of a swimmer's form. The use of the method is quite important and practically significant as several factors influence the preparation of the top form for the main swimming competition. The direct preparation period lasts for about three months, during which the competitor swims hundreds of kilometers on different intensities. Using the COMET, the values of attributes for each swimmer are introduced. The obtained model allows a broader analysis of the progress in terms of particular criteria sensitivity and robustness analysis.

Chapter "Brown-Gibson Model as a Multi-criteria Decision Analysis (MCDA) Method: Theoretical and Mathematical Formulations, Literature Review, and Applications" by Yimen et al. highlighted that different models of the MCDA method referred to as the Brown-Gibson model have so far been applied in various engineering and science fields, and different versions. The authors have provided a rich, complete and critical state-of-the-art literature survey of different versions. It includes theoretical and mathematical formulations of different versions, viz., original Brown-Gibson model, Buffa and Sarin version of Brown-Gibson model, the extended Brown-Gibson model, the Yimen & Dagbasi version of Brown-Gibson model, analytical hierarchy process (AHP)-integrated Brown-Gibson model and fuzzy Brown-Gibson model. An illustrative application of the original Brown-Gibson model in determining the optimal location to set a commercial center in Cameroon is provided with the choice associated with certain critical factors, objective factors and subjective factors. The authors highlighted that two facts confirm the importance of the Brown-Gibson model, viz., the model has seen several developments and applications since its inception and the model has a unique ability of combining the objective and subjective factors in decision-making.

In chapter "A Grey Approach for the Computation of Interactions Between Two Groups of Irrelevant Variables of Decision Matrices" Zakeri et al. emphasized that the existing MCDM methods merely provide solutions for the one-stage decision-making procedure and do not take other effective variables outside of the decision matrix into account, while in real-world processes, the decisions always have an impact on the variables where they appear to be irrelevant. The chapter aims at finding a mathematical solution to compute the impact between two irrelevant decision matrices in a complex decision-making problem using MCDM methods. The proposed strategies interaction model (SIM) approach is applied to a case of supplier selection and the strategies of the firm in which the interaction of selected strategies is investigated on the selection of the best supplier. The inherent uncertainty is handled using a four-section approach implemented as a grey framework and deals with grey Entropy, grey TOPSIS and the grey strategies interaction model.

Chapter "Statistical Analysis of KMM Program—An Educational Intervention" by Vaidya et al. highlighted that the educational interventions are intended to help struggling students by addressing their behavioral issues and social skills. Several criteria and factors need to be considered. In this regard, the chapter presents the complete life cycle of the intervention process implemented by the Keep Moving

Movement (KMM) as a pilot study. The impact of the KMM programme is analysed using correlation analysis, factor analysis and paired t-test. The group-wise and student-wise analysis of students reveals significant positive changes in positive thinking and willingness. The study provides possible extension towards a larger set of students to improve positive thinking, confidence and willingness.

Pune, India Anand J. Kulkarni

Acknowledgements

The author is grateful to the reviewers of the book for their valuable time and efforts in critically reviewing the chapters. Their critical and constructive reviews have certainly helped in the enrichment of every chapter. The editor would like to thank Ms. Kamiya Khattar for the editorial assistance and Rini Christy Xavier Rajasekaran for the book production assistance. The author hopes that the readers will find this book useful and valuable to their research.

Contents

Editor and Contributors

About the Editor

Anand J. Kulkarni holds a Ph.D. in Distributed Optimization from Nanyang Technological University, Singapore, MS in Artificial Intelligence from the University of Regina, Canada, Bachelor of Engineering from Shivaji University, India, and Diploma from the MSBTE, Mumbai. He worked as a Post Doctorate Research Fellow at Odette School of Business, University of Windsor, Canada. Dr. Kulkarni has worked with Symbiosis International University, Pune, India for over six years. Currently, he is a Professor & Associate Director at the Institute of AI at MITWPU. His research interests include optimization algorithms, multi-agent systems, complex systems, swarm optimization, and self-organizing systems. Anand pioneered socio-inspired optimization methodologies such as Cohort Intelligence, Ideology Algorithm, Expectation Algorithm, and Socio Evolution & Learning Optimization Algorithm. He is the founder and chairman of Optimization and Agent Technology Research Lab and has over 70 research papers in journals and conferences, 04 authored and 08 edited books to his credit. Dr. Kulkarni is the lead editor for the Springer and Taylor and Francis book series. He regularly writes on Artificial Intelligence in several newspapers and magazines. Dr. Kulkarni has delivered expert research talks in many countries such as the USA, Canada, Singapore, Malaysia, India, and France.

Contributors

Yungho Adamu The Higher Institute of Transport and Logistics (HITL), University of Bamenda, Bamenda, Cameroon

Syed Naseem Ahmad Department of Electronics and Communication Engineering, Jamia Millia Islamia (A Central University), New Delhi, India

Fereshteh Sattari Barabadi Department of Public Administration, Financial Management, Islamic Azad University, Mashhad, Iran

Divyajyoti Biswal Department of Chemical Engineering, Visvesvaraya National Institute of Technology, Nagpur, India

Marko Bohanec Department of Knowledge Technologies, Jožef Stefan Institute, Ljubljana, Slovenia;
University of Nova Gorica, Nova Gorica, Slovenia

Naoufel Cheikhrouhou Geneva School of Business Administration, University of Applied Sciences Western Switzerland, Geneva, Switzerland

Mustafa Dagbasi Department of Energy Systems Engineering, Cyprus International University, Nicosia, Cyprus

Fombe Lawrence Fon The Higher Institute of Transport and Logistics (HITL), University of Bamenda, Bamenda, Cameroon

Gholamreza Haseli Faculty of Industrial Engineering and Management, Shahrood University of Technology, Shahrood, Iran

Saurabh N. Joglekar Department of Chemical Engineering, Laxminarayan Institute of Technology, R.T.M. Nagpur University, Nagpur, India

Aleksandra Kaczyńska West Pomeranian University of Technology in Szczecin, Szczecin, Poland

Abraham Kanmogne National Advanced School of Engineering, University of Yaoundé I, Yaoundé, Cameroon

Zishan Husain Khan Department of Applied Sciences and Humanities, Jamia Millia Islamia (A Central University), New Delhi, India

Dimitri Konstantas Geneva School of Economics and Management, University of Geneva, Geneva, Switzerland

Giannis Kourousias School of Electrical and Computer Engineering, National Technical University of Athens, Zografou, Athens, Greece

Aradhana Kulkarni Consultant Chezuba, Corporate Social Responsibility Consultant Pune, Pune, India

Sachin A. Mandavgane Department of Chemical Engineering, Visvesvaraya National Institute of Technology, Nagpur, India

Shilpa Bhaskar Mujumdar Symbiosis Institute of Computer Studies and Research (SICSR) Symbiosis International (Deemed University) (SIU), Pune, Maharashtra, India

Mohd. Nazim Computer Science and Technology Research Group, Department of Applied Sciences and Humanities, Faculty of Engineering and Technology, Jamia Millia Islamia, (A Central University), New Delhi, India

Azra Parveen Department of Applied Sciences and Humanities, Jamia Millia Islamia (A Central University), New Delhi, India

Mohd. Sadiq Department of Computer Science and Automation, Indian Institute of Science, Bangalore, India

Wojciech Sałabun West Pomeranian University of Technology in Szczecin, Szczecin, Poland

Reza Sheikh Faculty of Industrial Engineering and Management, Shahrood University of Technology, Shahrood, Iran

Andrii Shekhovtsov West Pomeranian University of Technology in Szczecin, Szczecin, Poland

Shailaja Shirwaikar Department of Mathematics Savitribai, Phule Pune University, Pune, India

Eleftherios Siskos School of Electrical and Computer Engineering, National Technical University of Athens, Zografou, Athens, Greece

Yannis Siskos Department of Informatics, University of Piraeus, Piraeus, Greece

Theodore Tchotang National Advanced School of Engineering, University of Yaoundé I, Yaoundé, Cameroon

Anagha Vijay Vaidya Symbiosis Institute of Computer Studies and Research (SICSR) Symbiosis International (Deemed University) (SIU), Pune, Maharashtra, India

Chaudhary Wali Mohammad Computer Science and Technology Research Group, Department of Applied Sciences and Humanities, Faculty of Engineering and Technology, Jamia Millia Islamia, (A Central University), New Delhi, India

Jarosław Watróbski University of Szczecin, Szczecin, Poland

Jakub Więckowski West Pomeranian University of Technology in Szczecin, Szczecin, Poland

Nasser Yimen National Advanced School of Engineering, University of Yaoundé I, Yaoundé, Cameroon;
The Higher Institute of Transport and Logistics (HITL), University of Bamenda, Bamenda, Cameroon

Shervin Zakeri Geneva School of Economics and Management, University of Geneva, Geneva, Switzerland;
Geneva School of Business Administration, University of Applied Sciences Western Switzerland, Geneva, Switzerland

Abbreviations

3D	Three-dimensional
AHP	Analytic Hierarchy Process
AIFD	Active Image Forgery Detection
AMT	Advanced Manufacturing Technology
ANP	Analytic Network Process
APM	Assessment Preference Measure
AQ	Algorithm Quasi-optimal, a machine rule learning algorithm
ARP	Average Range of preferential Parameters
ARRI	Average Range of the Ranking Index
AS/RS	Automatic Storage and Retrieval System
ASI	Average Stability Index
AVG	Average
BPE	Biomass Processing Enterprise
BREEAM	Building Research Establishment Environmental Assessment Method
C	Criteria
CC	Closeness Coefficients
CDPC	Consistency-Driven Pairwise Comparisons, an MCDM method
CFI	Critical Factor Index
CFM	Critical Factor Measure
COMET	Characteristic Objects Method
COPRAS	Complex Proportional Assessment
COs	Characteristic Objects
CSM	Computer Software Measure
CT	Conventional Technology
CTE	Cost and Time Effectiveness
DECMAK	DECision MAKing, an early predecessor of DEX
DEX	Decision EXpert, a qualitative MCDM method
DEXi	Software implementing the DEX method
DIF	Digital Image Forensic
DM	Decision Maker

DRSA	Dominance-based Rough Set Analysis, an MCDM approach
DSS	Decision Support System
EBG	Extended Brown-Gibson
EC	Effective Cost
EEA	European Environmental Agency
ELA	Error Level Analysis
ELECTRE	ÉLimination et Choix Traduisant la REalité, a family of MCDM methods
ENF	Effective Non-Financial
ERA	Extreme Ranking Analysis
ET	Effective Time
F	Fair
FDM	Fuzzy Decision Matrix
FLSI	Fuzzy Location Selection Index
FNIS	Fuzzy Negative Ideal Solution
FOFC	Fuzzy Objective Factor of Alternative
FOFM	Fuzzy Objective Factor Measure
FPIS	Fuzzy Positive Ideal Solution
FR	Functional Requirement
FSFM	Fuzzy Subjective Factor Measure
FST	Fuzzy Set Theory
FUSE	Fuzzy Synthetic Extent
G	Good
GDP	Gross Domestic Product
GRIHA	Green Rating for Integrated Habitat Assessment
G-TOPSIS	Grey-TOPSIS
GUV	the Grey Uncertainty Value
H	High
HINT	Hierarchical INduction Tool, a machine-learning method for developing DEX models from data
ICT	Information and Communications Technology
IEC	Ineffective Cost
IES	Institute Examination System
IET	Ineffective Time
INF	Ineffective Non-Financial
JPEG	Joint Photographic Expert Group
L	Low
LCA	Life cycle assessment
LEED	Leadership in Energy and Environmental Design
LM	Location Measure
LP	Linear Programming
LPM	Location Preference Measure
M	Medium
MA	Metadata

MACBETH	Measuring Attractiveness by a Categorical Based Evaluation Technique, an MCDM method
MADA	Multi-Attribute Decision Analysis
MCDA	Multi-Criteria Decision Analysis
MCDM	Multi-criteria Decision Making
MCHP	Multi-Criteria Hierarchy Process, a hierarchical MCDM approach
MG	Medium Good
MIVES	Modelo Integrado de Valor para una Evaluación Sostenible
mmol	Millimole
MODA	Multi-Objective Decision Analysis
MP	Medium Poor
MSPM	Manufacturing System Preference Measure
NFR	Non-functional Requirement
NIS	Negative Ideal Solution
NPV	Net Present Value
OFC	Objective Factor Cost
OFM	Objective Factor Measure
P	Poor
PIFD	Passive Image Forgery Detection
PIS	Positive Ideal Solution
PROMETHEE	Preference Ranking Organization Method for Enrichment of Evaluations
PSPM	Pumping System Preference Measure
QFD	Quality Function Deployment
QQ	Qualitative-Quantitative, an approach to ranking of alternatives using a DEX model
QSPM	Quantitative Strategic Planning Matrix
R&D	Research and Development
RARR	Ratio of the Average Range of the Ranking
RMS	Reconfigurable Manufacturing Systems
ROR	Robust Ordinal Regression
S	Strong
SDGs	Sustainability Development Goals
SFM	Subjective Factor Measure
SIM	Strategies interaction model
SO	Strengths Opportunities
SPRI	Statistical Preference Relations Index
SR	Software Requirement
ST	Strengths Threats
SWOT	Strengths, Weaknesses, Opportunities, Threats
T	Tools
TFN	Triangular Fuzzy Number
TFNs	Triangular Fuzzy Numbers
TOPSIS	The Technique for Order of Preference by Similarity to Ideal Solution

TS	Traditional System
TVM	Time Value of Money
UTA	UTilités Additives
VG	Very Good
VH	Very High
VIKOR	VlseKriterijumska Optimizacija I Kompromisno Resenje
VL	Very Low
VP	Very Poor
VS	Very Strong
VW	Very Weak
W	The subjective factor weight
WN	Weighted Normalized Fuzzy Decision Matrix
WO	Weaknesses Opportunities
WT	Weaknesses Threats

Symbols

$\mathrm{X_{min}}$	Indicator value generating minimum satisfaction value
$\mathrm{X_{max}}$	Indicator value associated to maximum satisfaction
$\mathrm{X_{ind}}$	Value of indicator for which value function is to be calculated
$X_{n \times m}$	Decision matrix, where n stands for the number of alternatives and m for the number of criteria
x_{ij}	Element of the decision matrix, where $i = 1, 2, \ldots, n$ is an index of alternative and $j = 1, 2, \ldots, m$ is an index of criteria
r_{ij}	Element of the normalized decision matrix, where $i = 1, 2, \ldots, n$ is an index of alternative and $j = 1, 2, \ldots, m$ is an index of criteria
V	Value function
$\mathrm{P_i}$	Shape factor that is decided on the shape of the curve
$\mathrm{C_i}$	Axis value of the point where the curve changes its direction
$\mathrm{K_i}$	Response value to the Ci
A	A constant, usually considered to be 0
B	A constant, limits the function to a range of 0 to 1
$\mathcal{A} = \{A_1, A_2, \ldots, A_q\}$	Alternatives
$A_i \in \mathcal{A}, A_i = \{a_{x,i} \in E_x, \forall x \in X\}$	An alternative
$a_{x,i} \in A_i$	Value of A_i assigned to attribute x
$B_x \subset D_x$	Subset of bad values of attribute x
$C_y = \prod_{x \in S(y)} D_x$	Domain of f_y
D_x	Value scale of attribute x

E_x	Range of values that can be assigned to attribute x
E_y	Range of f_y
$e = (\mathbf{x}, y) \in T_y$	Elementary decision rule, an entry in T_y
F	Set of aggregation functions of a DEX model
$\mathcal{F}_y$	Fuzzy distributions over D_y
$f_y : C_y \to E_y$	Aggregation function associated with attribute y
g_y	An approximation of f_y
$G_x \subset D_x$	Subset of good values of attribute x
$I_i \subset A_i$	Subset of values of A_i, assigned to input attributes
I_y	Set of intervals over D_y
$m_x = \|D_x\|$	Number of categories of scale D_x
$M = (X, D, S, F)$	A DEX model
$N_x \subset D_x$	Subset of neutral values of attribute x
$O_i \subset A_i$	Subset of values of A_i, assigned to output (aggregate) attributes
$\mathrm{ord}(v_{x,i}) = i$	Ordinal value of $v_{x,i}$
$P(x)$	Set of parents of attribute x
$\mathcal{P}_y$	Probability distributions over D_y
$r_y = \|C_y\|$	Size of C_y and the corresponding T_y
$S : X \to 2^x$	Descendant function
$S(x)$	Set of descendants of attribute x
$\mathcal{S}_y$	The power set of D_y
$T_y = \{(\mathbf{x}, y), \mathbf{x} \in C_y, y \in E_y\}$	Decision table associated with attribute y
$v_{x,i} \in D_x$	i-th qualitative value (category) of attribute x
$v_{x,i} \preccurlyeq v_{x,j}$	Weak preference relation
$w, w_i \in \mathcal{R}$	Relative weight (importance) of an attribute
$x, x_i, y \in X$	An attribute
$X = \{x_1, x_2, \ldots, x_n\}$	Set of attributes
$\omega \in [-0.5, +0.5]$	An offset to qualitative value v
$\sum$	Summation
$\otimes$	Fuzzy multiplication operator
min ( )	Returns minimum value
max ( )	Returns maximum value
α	The objective factor decision weight
$\mu_{\tilde{A}}$	is the value of the membership function
C_r	set of the fuzzy numbers, where C_i represents each criterion
CO	Characteristic Objects as Cartesian Product
a_{ij}	element of the matrix of expert judgement, where $i = 1, 2, \ldots, n$ is and index of row in matrix and $j = 1, 2, \ldots, m$ is an index of matrix' column

SJ_i	is the sum of thc *i-th* row of matrix of expert judgement
y	is the value of linear function
β_0	is the coefficient of the independent variable in linear function
β_1	is a directional coefficient in linear function
x	is the independent variable in linear function
$\otimes G_1$	A grey number
$\left[\underline{G_1}, \overline{G_1}\right]$	Grey interval
$\underline{G_1}$	Grey lower bound
$\overline{\overline{G_1}}$	Grey upper bound
e	Entropy
w	Weight
S^{max}	Positive ideal alternative
S^{min}	Negative ideal alternative
γ_{oi}	The grey relation coefficient
C_i	The grade of grey relation
$\otimes P = \left[\underline{P_{ij}}, \overline{P_{ij}}\right]$	A normalized grey number
$N_D = \left[N_{\underline{G}_{ij}}, N_{\overline{G}_{ij}}\right]$	The normalized decision matrix
D	The decision matrix
ℓ	The distance between the elements of each cloud with lower and upper bound
ϱ	GUV
ξ	The coefficient of uncertainty/probability

MIVES: A Multi-Attribute Value Function-Based Methodology for Sustainability Assessment

Divyajyoti Biswal, Saurabh N. Joglekar, and Sachin A. Mandavgane

Abstract Sustainability assessment remains an important aspect of any process or product design and is dependent on the triple bottom approach of economic, environment and social criteria. However, the subjectivity associated with it makes the results of the assessment difficult to comprehend. MIVES—a multi-attribute value function-based methodology—provides the necessary framework for sustainability assessments. The methodology compares the alternative with respect to other available alternatives or arbitrary standards. The value function for each indicator is evaluated (between 0 and 1) based on the maximum (value 1) and minimum (Value 0) desired value of the indicators. The overall value function is then calculated based on the weightage of each indicator and criteria. Thus the overall value function or "Sustainability index" takes into account the different aspects of the alternatives considered for evaluation. The sustainability index nearing 1 represents greater sustainability and vice versa. The framework assigns a number to each alternative based on the indicator values, weightage of indicators and most and least desired conditions. With this perspective, the chapter discusses different case studies such as construction, aviation, education and process industries where this methodology is applied to come up with a logical and more comprehensive sustainability evaluation framework.

Keywords Sustainability assessment · Multi-criteria decision-making · Sustainability index · Framework

D. Biswal · S. A. Mandavgane (✉)
Department of Chemical Engineering, Visvesvaraya National Institute of Technology, Nagpur, India
e-mail: sam@che.vnit.ac.in

S. N. Joglekar
Department of Chemical Engineering, Laxminarayan Institute of Technology, R.T.M. Nagpur University, Nagpur, India

A. J. Kulkarni (ed.), *Multiple Criteria Decision Making*, Studies in Systems, Decision and Control 407, https://doi.org/10.1007/978-981-16-7414-3_1

Abbreviations

AHP	Analytical Hierarchy Process
BPE	Biomass Processing Enterprise
BREEAM	Building Research Establishment Environmental Assessment Method
EEA	European Environmental Agency
LCA	Life cycle assessment
GRIHA	Green Rating for Integrated Habitat Assessment
LEED	Leadership in Energy and Environmental Design
MIVES	Modelo Integrado de Valor para una Evaluación Sostenible
SDGs	Sustainability Development Goals

List of Symbols

X_{min}	Indicator value generating minimum satisfaction value
X_{max}	Indicator value associated to maximum satisfaction
X_{ind}	Value of indicator for which value function is to be calculated
V	Value function
P_i	Shape factor that is decided on the shape of the curve
C_i	Axis value of the point where the curve changes its direction
K_i	Response value to the C_i
A	A constant, usually considered to be 0
B	A constant, limits the function to a range of 0 to 1

1 Introduction

Sustainability assessment is gaining importance in several areas of research. The implementation of ambitious sustainable development goals (SDGs) has enabled the industry to think of sustainability not as a mere compliance but also as a tool to plan growth. Sustainability assessment is a multi-dimensional subjective concept and hence there are many attempts to develop a framework that addresses the subjectivity involved. Multi-criteria decision-making methods show great promise in the integration of different aspects involved in overall sustainability assessment.

The majority of the sustainability assessment frameworks such as BREAM, LCA, GRIHA, EcoEffect, Green Star, LEED, etc. have been developed focusing on the construction sector. Most of the newly proposed frameworks address more than one sustainability requirement such as technical, functional and governance. With a view to addressing the subjectivity involved in overall sustainability assessment, MIVES methodology was developed for industrial buildings [1]. The chapter intends to explore MIVES (Modelo Integrado de Valor para una Evaluación Sostenible)-integrated multi-criteria decision-making methods applied to processes. MIVES—a tool developed by researchers from three Spanish universities and institutes (UPC, UPV and Labein-Tecnalia)—was initially applied in the field of sustainability and industrial buildings. MIVES integrates the multi-criteria decision method and the multi-attribute utility theory through the incorporation of the value function concept to the different criteria. The criteria are assigned weightages through an analytical hierarchy process. Analytical hierarchy process (AHP)—originally developed by Saaty [2]—is a linear additive model that assigns scores or weightages to different criteria based on their relative importance. The model is based on pairwise comparisons of the criteria/indicators.

MIVES approach prescribes a common equation that evaluates the alternatives based on the decision-maker's preference and presents the results in a comprehensive and easy way. The value function takes into account the physically meaningful variables and allows a common platform for their comparison. In short, the value function converts the qualitative and quantitative indicators of particular criteria on a scale of 0 (least desirable) and 1 (most desirable) [3]. The value function for the most desirable indicator value is assigned value 1, whereas the least desirable indicator value is assigned value 0. The alternative's indicator value function is defined based on this scale of 0–1. The value function is primarily dependent on the following variables—tendency of function (increasing or decreasing), nature of curve (linear, S-shaped, convex or concave), the variable value, shape factors and minimum and maximum satisfaction values. The maximum and minimum satisfaction values can be either among the available alternatives or they can be arbitrarily decided based on the industry/societal/economic/product standards as well.

The fundamental definition of sustainable development was published in the Brundtland Commission of 1987 that stated the development to meet the present needs without compromising the future needs. Since then there are various attempts to define sustainable development. One such school of thought represents sustainable development to be based on the triple bottom line concept. That implies sustainability assessment be based on economic, environment and social criteria having equal importance in decision-making [4].

Hence MIVES methodology with its ability to incorporate different variables in one common framework can play a significant role in evaluating the overall sustainability of any product, process and service. MIVES though initially was developed to compare different products or services related to the construction industry, the methodology can be applied to other sectors such as chemical processes as well. The framework is dependent on the indicators and their weightage, hence caution has to be exercised while deciding on the indicators and weightages. The number of indicators describing the alternative is also of prime importance. The number of indicators should be based on "necessary and sufficient" conditions to describe the alternative. A high number of indicators would result in decreased sensitivity of the framework as the relative weightage would be distributed among the alternatives.

A broad outline of the methodology can be explained in Fig. 1.

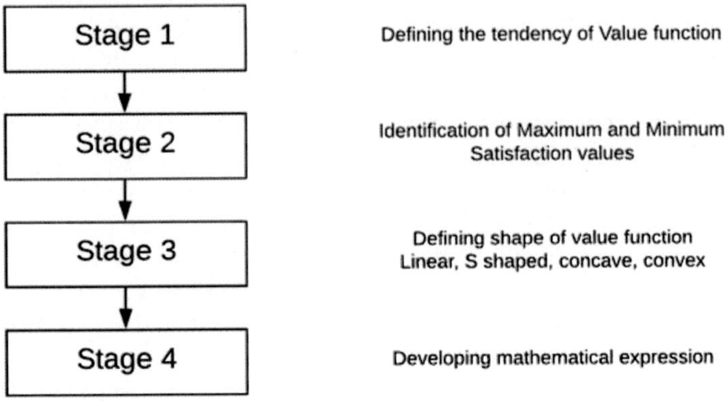

Fig. 1 Different stages of MIVES methodology

2 Methodology

Stage 1: The first step in adopting the methodology is to decide the tendency of the curve. If the satisfaction increases with an increase in the indicator value in such a case an increasing function is used. If the satisfaction decreases with a decrease or increase in indicator values, a decreasing function is adopted. The example of increasing function can be margins offered; percentage of renewable energy utilized etc. The example of decreasing function can be hazardous byproduct generation, environmental impacts etc.

Stage 2: Identification of the indicator values that generate the minimum and maximum satisfaction values is necessary. The methodology compares the present alternative based on the minimum and maximum desired value. X_{min} is an indicator value-generating minimum satisfaction value; X_{max} is an indicator value associated with maximum satisfaction. The minimum satisfaction condition can be the indicator among the alternatives or it can be decided based on presiding rules and regulations pertaining to the alternatives, with prior experiences of the operator. The range between the minimum and maximum desired conditions is not necessarily decided by the indicators of alternatives available.

Stage 3: Post decision on the maximum and minimum desired values, the type of curve is decided. The curves can be of four types—concave, convex, linear and S-shaped. Table 1 shows the reasoning for choosing a particular type of curve.

Stage 4: The value functions are calculated based on the general exponential equation Eq. (1), as mentioned below [5]:

$$V = A + B\left[1 - e^{-K_i^{(|X_{ind} - X_{\min}|/C_i)P_i}}\right] \tag{1}$$

Generally A = 0.

Table 1 Types of curve, characteristic and conditions of use

Type of curve	Reason	Condition of use
Concave	Satisfaction increases at a higher rate with respect to an increase in indicator values from the point of minimum satisfaction	A change in point generating minimum satisfaction is highly valued
Convex	Satisfaction hardly increases with respect to an increase in indicator values from the point of minimum satisfaction	It is more important to reach the maximum satisfaction value
Linear	Linear or proportional relationship	Can be used as default
S-shaped	A significant change in satisfaction at central values	The slope of satisfaction with respect to indicator value is flat near to point of minimum and maximum satisfaction values

P_i is the shape factor that is decided on the shape of the curve.

X_{ind} is the value of the indicator for which the value function is to be calculated.

C_i is the axis value of the point where the curve changes its direction.

K_i is the response value to the C_i.

B limits the function to a range of 0 and 1.

$$B = \frac{1}{1 - e^{-K_i(|X_{\max} - X_{\min}|/C_i)^{P_i}}} \tag{2}$$

It is seen from Eq. (1) that the shape of the value function is strongly dependent on the values of C, K and P. The conditions for selecting values of C, K and P are shown in Table 2.

3 Application of MIVES in Different Sectors

3.1 *MIVES in Aviation Industry*

The application of MIVES can be observed in a plethora of industries and sectors. A case study done on El Prat Airport, Spain explains the application of MIVES coupled with life cycle assessment for sustainability evaluation of ground-handling operations at the location. The data collected in real time encompasses airport ground operations over a span of two weeks, one to represent regular traffic and another to mimic holiday gridlock. The environmental impact due to landing and takeoff (LTO) activities in the airport was quantified using the tool "Master Emission Calculator for Aviation",

Table 2 Conditions for selecting values of *C*, *K* and *P*

Increasing function			
Function	C	K	P
Linear	$C \approx X_{min}$	≈ 0	≈ 1
Convex	$X_{min} + \frac{X_{max}-X_{min}}{2} < C < X_{min}$	< 0.5	> 1
Concave	$X_{min} < C < X_{min} + \frac{X_{max}-X_{min}}{2}$	> 0.5	< 1
S-shaped	$X_{min} + \frac{X_{max}-X_{min}}{5} < C < X_{min} + \frac{4(X_{max}-X_{min})}{5}$	0.2/0.8	> 1
Decreasing function			
Linear	$C \approx X_{min}$	$\approx A0$	≈ 1
Convex	$X_{max} < C < X_{max} + \frac{X_{min}-X_{max}}{2}$	< 0.5	> 1
Concave	$X_{min} - \frac{X_{min}-X_{max}}{2} < C < X_{min}$	> 0.5	< 1
S-shaped	$X_{max} - \frac{4(X_{max}-X_{min})}{5} < C < X_{max} - \frac{X_{max}-X_{min}}{5}$	0.2/0.8	> 1

developed by the European Environmental Agency (EEA) while SIMAPRO was used for LCA. MIVES methodology was used to determine the sensitivity to change of desired values of parameters set for evaluation in two different scenarios. The first corresponds to the use of a 100% combustion engine vehicle fleet while the second enlists a net 20% substitution of the ground vehicles fleet by electricity-powered counterparts [6].

3.2 MIVES in Construction and Architecture

Alberti et al. [7] reported a specific case study associated with the construction sector. The study considered differing reinforcements used in concrete slabs used for the construction of the La Canda Tunnels, where some concrete slabs were reinforced with conventional steel mesh while the others with polyolefin [7]. Concrete is put together by mixing cement, water and, in some cases, chemical additives for suitability and pliability. The variety in the production of different types of cement decides the strength and other parameters for concrete. Infrastructure development inevitably requires the application of concrete for structural purposes. Concrete is typically reinforced with steel for increased strength and flexibility. Depending on the load-bearing conditions specified for the construction, conventional steel of different dimensions may be used. However, the research in the field of alternate construction materials has rapidly increased in recent times due to the current emphasis on the sustainability aspect. Alberti et al. [7] discuss the sustainability assessments carried out for one such alternate construction reinforcement material, polyolefin, in comparison with the conventional steel augmentation. Such raw materials used for reinforcement for the purpose of building green construction materials are mostly used in non-structural applications. The study analyzes the usage of concrete slabs separately reinforced with conventional steel mesh and polyolefin fibers for handling equal amounts of load across their service life, thereby creating an immaculate comparison between the two. Environmental, social and economic parameters were evaluated for the determination of the overall sustainability scores of the two alternatives. The steel mesh consisted of B500S steel bars (dimensions: 150 × 150 × 6 mm) while commercial fibers (SikaFibre T48) were used as polyolefin supplements. The overall better alternative turned out to be the conventional steel mesh with a sustainability score of 74 out of 100 [7]. Table 3 enumerates the different indicators used for the sustainability evaluation and their corresponding category-wise scores.

Pons et al. [1] have discussed the use of MIVES as a sustainability assessment method in civil engineering and architecture [1]. The article discusses the application of MIVES for the sustainability assessment of different technologies used in building schools [8] and an array of methods for manufacturing structural concrete columns [9]. Pons and Aguado [8] analyzed a diversity of technologies used for the construction of academic edifices, which include the following construction systems—onsite concrete system, prefabricated concrete framed structure, prefabricated steel modular arrangement and prefabricated timber structure system. The latter is considered only because of its superlative environmental performance, even though in practice they are seldom used [10]. The requirement tree for the analysis of the alternatives was constructed with an emphasis on economic indicators (50% weightage) while environmental and social indicators were given 30% and 20% weightages, respectively. Subsequent calculations following the standard procedure revealed that the prefabricated concrete framed structure systems turned out to have the highest sustainability index, i.e., 0.72 on a scale of 0 to 1. Prefabricated steel modules structure system also secured a score in the same neighborhood (0.71 on a scale of 0 to 1). However, the onsite concrete structure system received the lowest sustainability score (0.35). Figure 2 presents the comparison made on the basis of sustainability indices as obtained from the analysis.

The article by Pons et al. [1] also discusses the study done by Pons and Fuente in 2013 that discussed sustainability index evaluation of structural concrete columns divided across three baselines—cross-section shape (square or circular), characteristic concrete strength (compressive strength of cylindrical cross-section at 28 days) and manner of compacting of concrete (self-compacting or vibrating). These divisions were studied across 12 different sample structures. Ten different indicators spread across the categories of social, economic and environmental aspects were evaluated for the determination of the overall sustainability index. The study was conducted with 50% of weightage given to economic preconditions while environmental and social factors were weighed in at 33% and 17%, respectively. The behavior of each indicator was evaluated through the value functions for their respective parameters. The study concluded with the circular cross-sectioned self-compacting concrete pillar with 30 cm diameter as the best alternative of the 12 available options (Table 4). The following observations were also made:

- Greater strength concrete have higher sustainability
- Self-compacting mode preferred over vibrational packing

Table 3 Indicator list and category-wise sustainability scores—comparison between steel mesh and polyolefin reinforcement [7]

Requirement	Indicator name	Indicator score		Overall score	
		Polyolefin fibers	Steel mesh	Polyolefin fibers	Steel mesh
Economic (50%)	Total costs (inclusive of construction time)	71.67	72.00	22.47	40.03
	Non-quality costs	45.50	72.00		
	Dismantling costs	30.00	80.00		
	Cost of service, maintenance, energy, change of use	8.48	76.80		
	Resilience, risk of disaster × cost of reconstruction + lack of use	13.33	13.33		
Environmental (30%)	Cement, construction	3.00	3.00	13.87	8.37
	Aggregates	3.17	3.17		
	Reinforcement (steel mesh, steel fibers, polyolefin fibers), construction	2.40	9.30		
	Water, construction	4.03	4.03		
	Auxiliary materials, construction	3.75	3.75		
	Reused material, construction	0.00	0.00		
	Cement, maintenance	7.84	9.22		
	Aggregates, maintenance	0.98	4.92		
	Reinforcement (steel mesh, steel fibers, polyolefin fibers), maintenance	0.31	1.55		

(continued)

Table 3 (continued)

Requirement	Indicator name	Indicator score		Overall score	
		Polyolefin fibers	Steel mesh	Polyolefin fibers	Steel mesh
	Water, maintenance	1.15	5.76		
	Auxiliary materials, maintenance	0.63	3.13		
	Reused material, maintenance	0.00	0.00		
	Global warming potential, construction	19.12	42.40		
	Total waste, construction	5.98	13.25		
	Global warming potential, maintenance	19.12	42.40		
	Total waste, construction, maintenance	5.98	13.25		
	Embodied energy	20.00	20.00		
	Construction energy	40.00	40.00		
	Service and maintenance energy	8.00	12.00		
Social (20%)	Comfort, thermal, air, noise, etc.	10.00	10.00	20.00	13.20
	Noise pollution, construction	15.00	15.00		
	Particles pollution, construction	15.00	15.00		
	Traffic disturbances, construction	15.00	15.00		

(continued)

Table 3 (continued)

Requirement	Indicator name	Indicator score		Overall score	
		Polyolefin fibers	Steel mesh	Polyolefin fibers	Steel mesh
	Noise pollution maintenance	3.00	15.00		
	Particles pollution, maintenance	3.00	15.00		
	Traffic disturbances, maintenance	3.00	15.00		
	Health and safety during construction	40.00	40.00		
	Health and safety during maintenance	8.00	40.00		
	Occupant safety, risk of disaster × cost of life disruption	20.00	20.00		

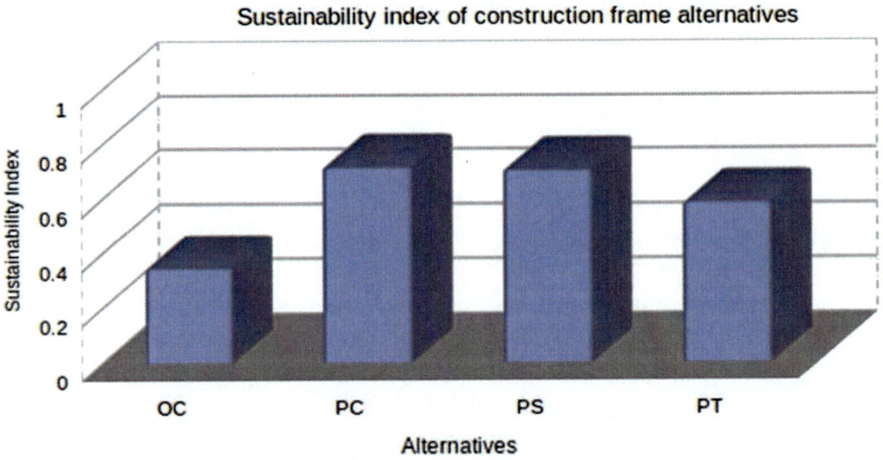

Fig. 2 Sustainability indices of construction frame alternatives [1], where OC—onsite concrete structure; PC—prefabricated concrete frame, PS—prefabricated steel modules, PT—prefabricated timber structure

Table 4 Sustainability indices of concrete pillar alternatives [9]

S. No.	Alternative	Cross-section type	Concrete characteristic strength	Compaction	Sustainability index
1	Sq1	Square	25	Self	0.61
2	Sq2	Square	25	Vibrated	0.62
3	Sq3	Square	50	Self	0.72
4	Sq4	Square	50	Vibrated	0.66
5	Sq5	Square	75	Self	0.77
6	Sq6	Square	75	Vibrated	0.71
7	Ci1	Circular	25	Self	0.56
8	Ci2	Circular	25	Vibrated	0.56
9	Ci3	Circular	50	Self	0.77
10	Ci4	Circular	50	Vibrated	0.72
11	Ci5	Circular	75	Self	0.85
12	Ci6	Circular	75	Vibrated	0.79

- Circular cross-section pillars are preferred when high performance concretes are used with lower cross-sectional area while square ones are preferred in case of higher cross-sections [9]

Another study conducted by Hosseini et al. in 2016 evaluates the alternatives proposed for the construction of temporary housing units (THUs) for the displaced population of the Bam earthquake (2003), a region in the Kerman province of Iran. The aftermath left 80% of buildings in Bam in ruins [11] and roughly 30% of the population was wiped out [12], leaving around 75,000 people homeless [13]. THUs are a temporary alternative to shelter people affected by natural disasters till permanent housing projects are complete. Often provided as prefabricated systems, THUs have long faced opposition on economic, environmental and social grounds despite being the default alternative while handling displaced populations. In this study, substitute technologies for manufacture/prefabrication of THUs as suggested by a private tech firm are evaluated with the objectives of determining the most sustainable alternative as well as testing the designed model. The alternatives included four variations in wall composition while sporting two different compositions for the roof (Fig. 3). The analysis indicated that concrete masonry units (CMU) were more sustainable than the others [14].

In another study, Joglekar et al. [15] reported the sustainability assessment of brickwork used for low-cost housing units. This study takes into account five specific alternatives to the production of bricks for low-cost housing units and evaluates the most sustainable option. Five different wall materials, namely clay bricks (CB), fly ash bricks (FB), cotton and paper waste bricks (CWB), paper and RHA waste bricks (RHAB) and paper mill waste bricks (PWB), were evaluated under social, economic, environmental and technical categories. Data for the same were generated

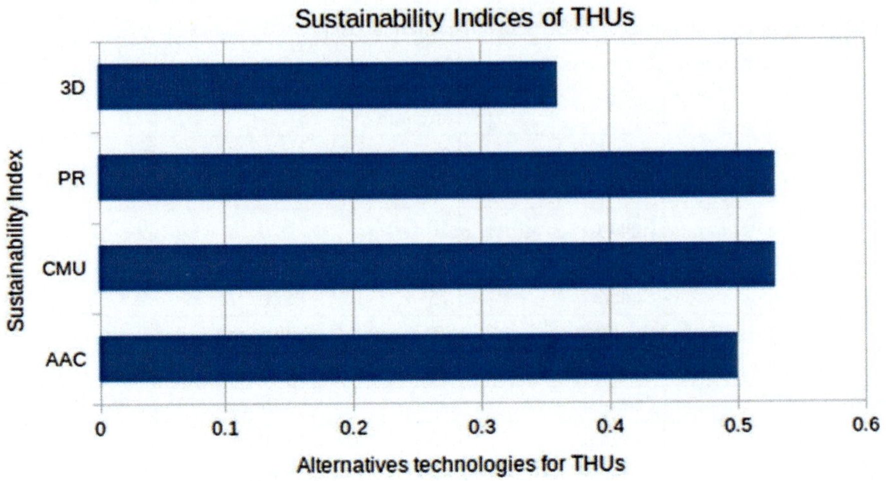

Fig. 3 Sustainability indices for alternative technologies used for construction of THUs in Bam, 2003 [14], where AAC—autoclaved aerated concrete blocks; CMU—concrete masonry units; PR—pressed reeds; 3D—3D sandwich panels

through literature assessment and process development (in case of unavailability of desired data). The system boundary of the study strictly included the life cycle phases of the production of bricks till the production stage. Apart from conventional indicators for social, economic and environmental index evaluation, specific technical indicators like thermal conductivity, compressive strength and bulk density were considered for appropriate evaluation of comparative assessment. Interestingly, in comparison with the conventional bricks used for the purpose, CWB turned out to have the highest sustainability index of 0.94 (on a scale of 0 to 1) with exceptional economic advantages while exhibiting comparable scores on environmental, social and technical fronts. Figure 4 shows the graphical representation of the results and contribution of each criterion to the overall sustainability index [15].

3.3 MIVES in Education and Teaching

The application of MIVES can also be observed in the education sector. A study conducted by Viñolas et al. (2009) discusses the application of multi-criteria decision-making methods for the evaluation of the work of professors in universities taking into account aspects of the profession. The work evaluation considers the assessment of professors on the basis of the investigative outlook, teaching ability, commitment to the system and extended work engagements. These requirements are divided across a plethora of indicators that carefully assess the performance of a professor in each category. The analysis was conducted on a relative scale, where the faculties were

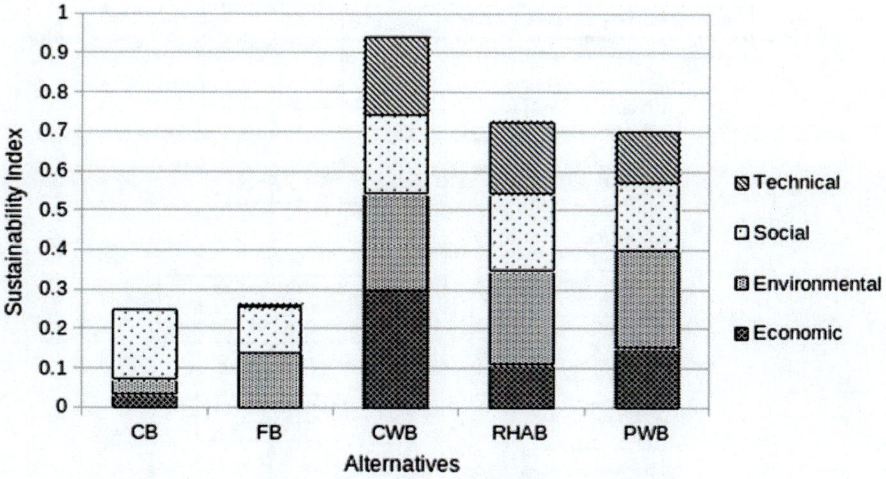

Fig. 4 Sustainability index for five alternative bricks for the brickwork of low-cost housing [15]

Table 5 Requirement category and respective criteria of evaluation used

S. No.	Requirement	Criteria
1	Investigation (35%)	Competitive projects with public resources
		Competitive projects with private resources
		Research results
2	Teaching (35%)	Teaching experience
		Teaching innovation
		Teaching results
3	Commitment to system (10%)	External to university
		Internal to university
4	Extended work engagements (20%)	Professional
		Social

weighed in comparison with one another. Table 5 enlists the requirement category and respective criteria of evaluation used in the study [16].

3.4 *MIVES in Enterprise—Case Study of Biomass Processing Enterprise*

Applying MIVES, Joglekar et al. (2019) have discussed the evaluation of the sustainability index of an existing biomass processing enterprise (BPE). The author also

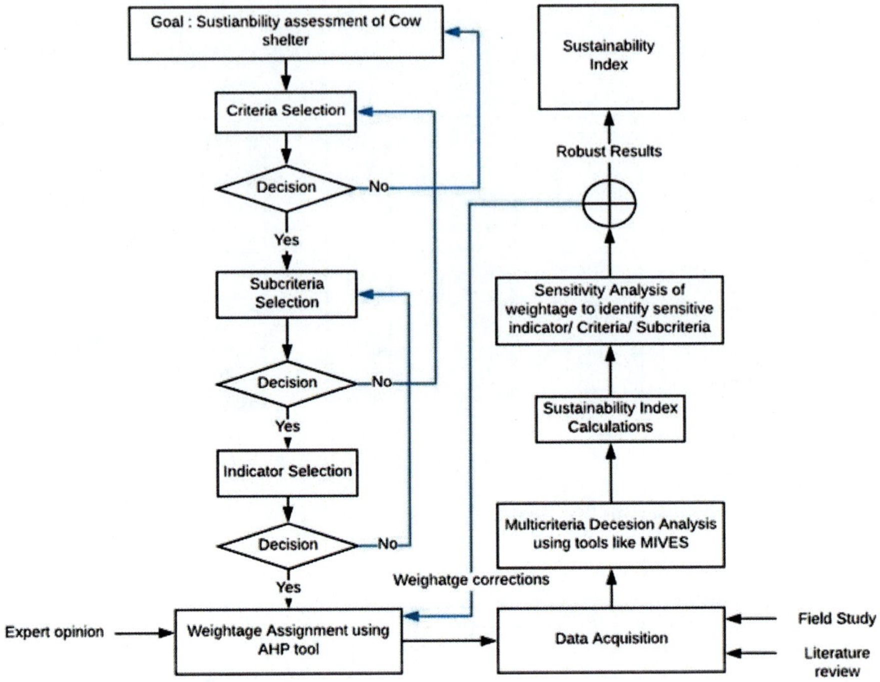

Fig. 5 General methodology to be adopted for sustainability evaluation of BPE [17]

prescribed the general framework to be adopted for applying MIVES to any enterprise (Fig. 5). The BPE is an establishment rearing breeds of cattle that use cow dung and urine as raw materials for the processing of products. The BPE considered in the study is representative of small-scale agricultural ventures that play vital roles in generating employment, mostly in the vicinity of the establishment, that are key to the socio-economic development of the region. Contrary to the other case studies discussed above that discuss the application of MIVES through a comparative approach, this study focuses on the evaluation of the sustainability index of the individual enterprise alone. The activities undertaken by the organization are evaluated under categories of social, economic and environmental criteria. The economic criteria are evaluated by gross profit earned by the BPE. As detailed inventory values were not available life cycle assessment was not performed and the environmental criteria are evaluated using indicators renewable energy used and product replacement potential. The social criteria are evaluated using indicators such as financial status, health impact, prerequisite skill requirement and social stature. The social criterion is evaluated by considering the feedback obtained from the workers of the BPE. The evaluation results in an overall sustainability index of 0.69 [17]. The application of MIVES for sustainability evaluation of enterprises can have an excellent reach, especially for social enterprises. The methodology can also be used by existing industries to benchmark their performance and can be used for comparison.

4 Conclusion

Sustainability evaluation has been gaining importance in the recent context, especially in chemical processing industries. Sustainability evaluations should include economic, environment, social, technical and ecological criteria/indicators. Thus a major obstacle of sustainability assessments is the subjectivity associated with these criteria/indicators. MIVES provides a solution to address the subjectivity, through which one can compare different alternatives product or process based on the maximum and minimum satisfaction values. Though the framework has been developed for the construction sector, the same can be applied with modifications to other domains as well. However, the basis of the framework stands on the pillars of overall weightage assigned to indicator/criteria and hence the results are entirely dependent on the same. With an increase in the number of indicators and/or criteria, the sensitivity of the results seems to reduce. Hence it is advisable to consider only the relevant indicators for analysis and improved dependence evaluation. The indicator, ideally, should take into account precise considerations depending on the different aspects of the criteria under which it is defined. Moreover, since the method uses constants (C, K and P) that are indicator-dependent, condition specifications for the same become utterly essential.

With this viewpoint, the chapter discusses the application of MIVES to different domains such as construction, aviation, education and process industry. The prescribed methodology is followed in each scenario with minor case-specific modifications. The methodology primarily rates the alternative based on its indicator values with respect to the least desired (value 0) and most desired value (value 1). The value function of the indicator is evaluated based on a generalized equation that compares the alternative with respect to other alternatives or defined limits. The overall value function generated for a specific alternative is dependent on the weightage of criteria/subcriteria/indicator, the value of the indicator and maximum and minimum desired conditions. Thus the overall function represents the sustainability rating or "Sustainability index" of the alternative for the specific evaluation procedure.

The methodology holds promise to address the need for a framework for sustainability evaluation in different domains of the industry.

Acknowledgments The authors are thankful to the Department of Science and Technology, Government of India for research funding (No. DST/SSTP/2018/76(G)).

References

1. Pons, O., De la Fuente, A., Aguado, A.: The use of MIVES as a sustainability assessment MCDM method for architecture and civil engineering applications. Sustainability **8**, 1–15 (2016). https://doi.org/10.3390/su8050460
2. Saaty, T.L.: The Analytic Hierarchy Process: Planning, Priority Setting, Resource Allocation: McGraw-Hill. New York (1980)
3. Pujadas, P., Pardo-Bosch, F., Aguado-Renter, A., Aguado, A.: Mives multi-criteria approach for the evaluation, prioritization, and selection of public investment projects. a case-study in the city of barcelona. Land Use Policy **1**, 29–37 (2017)
4. Pope, J., Annandale, D., Morrison-saunders, A.: Conceptualising sustainability assessment. Environ. Impact Assess. Rev. **24**, 595–616 (2004). https://doi.org/10.1016/j.eiar.2004.03.001
5. Alarcon, B., Aguado, A., Manga, R., Josa, A.: A value function for assessing sustainability: application to industrial buildings. Sustainability 35–50 (2011). https://doi.org/10.3390/su3010035
6. Miguel, P.Z. De: Sustainability analysis of the ground handling operations using MIVES methodology. Case study. El Prat airport. Universitat Politècnica de Catalunya (2017)
7. Alberti, M.G., Jaime, C.G., Enfedaque, A., Carmona, A., Valverde, C., Pardo, G.: Use of steel and polyolefin fibres in the La Canda tunnels: applying MIVES for assessing sustainability evaluation. Sustainability **10**, 1–11 (2018). https://doi.org/10.3390/su10124765
8. Pons, O., Aguado, A.: Integrated value model for sustainable assessment applied to technologies used to build schools in Catalonia, Spain. Build. Environ. **53**, 49–58 (2012). https://doi.org/10.1016/j.buildenv.2012.01.007
9. Pons, O., de la Fuente, A.: Integrated sustainability assessment method applied to structural concrete columns. Constr. Build. Mater. **49**, 882–893 (2013). https://doi.org/10.1016/j.conbuildmat.2013.09.009
10. Hormias Laperal, E., Bestraten Castells, S.: Cross-laminated timber construction - an introduction. AITIM Tech. Inf. Bull. (2009)
11. Havaii, M.H., Hosseini, M.: Bam Earthquake: From emergency response to reconstruction Mohammad. J. Sustainable Energy Environ. 229–237 (2004)
12. Kuwata, Y., Takada, S., Bastami, M.: Building damage and human casualties during the Bam-Iran earthquake. Asian J. Civil Eng. (Building Housing) **6**, 1–19 (2005)
13. Khazai, B., Eeri, M., Hausler, E., Eeri, M.: Intermediate shelters in Bam and permanent shelter reconstruction in villages following the 2003 Bam, Iran, Earthquake. Earthq. Spectra **21**, 487–511 (2005). https://doi.org/10.1193/1.2098907
14. Hosseini, S.M.A., De La Fuente, A., Pons, O.: Multi-criteria decision-making method for assessing the sustainability of post-disaster temporary housing units technologies: A case study in Bam, 2003. Sustainable Cities Soc. **20**, 38–51 (2016). https://doi.org/10.1016/j.scs.2015.09.012
15. Joglekar, S.N., Kharkar, R.A., Mandavgane, S.A., Kulkarni, B.D.: Sustainability assessment of brick work for low-cost housing: A comparison between waste based bricks and burnt clay bricks. Sustainable Cities Soc. **37**, 396–406 (2018). https://doi.org/10.1016/j.scs.2017.11.025
16. Viñolas Prat, B., Aguado de Cea, A., Josa Garcia-Tornel, A., Villegas Flores, N., Fernández Prada, M.Á.: Aplicación del análisis de valor para una evaluación integral y objetiva del profesorado universitario. Rev. Univ. y Soc. del Conoc. **6**, 22–37 (2009)
17. Joglekar, S.N., Darwai, V., Mandavgane, S.A.: A methodology of evaluating sustainability index of a biomass processing enterprise: a case study of native cow dung – urine biorefinery. Environ. Sci. Pollut. Res. **27**, 27435–27448 (2019)

Base Criterion Method (BCM)

Gholamreza Haseli and Reza Sheikh

Abstract The base criterion method (BCM) is one of the latest MCDM methods introduced to obtain the weight of the criteria. This method was introduced in 2020 by Haseli et al. The BCM method uses the pairwise comparison approach to obtain the weight of the criteria. This method removes a large number of unnecessary comparisons by dividing pairwise comparisons into two categories: base comparisons and final comparisons. To obtain the weight of the criteria with the BCM method, only base comparisons are needed. In the base comparisons for n criteria, n − 1 pairwise comparisons need to be performed. The results in the BCM method will be fully consistent because instead of the controlling outputs of the pairwise comparisons to measure the inconsistency, the BCM method controls the inputs of pairwise comparisons. By controlling the input values of pairwise comparisons, there will be no more errors in the process of obtaining weights. The zero error accuracy means optimal and full consistent weights. Therefore, the BCM method can obtain the optimal weight of the criteria quite accurately. Also, it is required to perform fewer pairwise comparisons compared to the other existing MCDM methods. In this chapter, examples are provided to become more familiar with the problem-solving process for obtaining the weight of criteria using the BCM method.

1 Introduction

Problems in the real world are very complex and it is impossible to make a decision just by considering one criterion [1]. Decision-making is a complex and difficult process because the decision-makers have to consider several factors at the same time. Therefore, to reach an optimal decision, identifying and evaluating all the criteria is necessary. Decision-making in which multiple criteria influence

G. Haseli (✉) · R. Sheikh
Faculty of Industrial Engineering and Management, Shahrood University of Technology, Shahrood, Iran

R. Sheikh
e-mail: resheikh@shahroodut.ac.ir

A. J. Kulkarni (ed.), *Multiple Criteria Decision Making*, Studies in Systems, Decision and Control 407, https://doi.org/10.1007/978-981-16-7414-3_2

the decision is known as multi-criteria decision-making (MCDM) [2]. In MCDM methods, decision-makers evaluate the different options based on the set of criteria and then choose the best one as a preferred option [3, 4]. Therefore, the MCDM methods have attracted a lot of attention in various scientific fields [5]. Based on the problem-solving space, MCDM can be divided into two categories: the multi-attribute decision-making (MADM) and multi-objective decision-making (MODM). The MADM first evaluates the different options, then listed the superior and inferior options, and then selects the desired option. Also, the MODM methods employ the vector-based optimization method, which is a type of mathematical programming method [2].

The MCDM methods are divided into two parts when dealing with practical problems. First, obtain decision information, including the weight of the criteria. Second, evaluate the options according to the criteria information [6]. In the last years, several methods have been introduced for obtaining the weight of criteria based on pairwise comparisons, such as the analytic hierarchy process (AHP) [7], analytic network process (ANP) [8], best-worst method (BWM) [6, 9] and finally base criterion method (BCM) [10, 11].

The AHP is an important method that used pairwise comparisons for building preferential relationships between the criteria [12]. In this method, the factors are sorted in a hierarchical structure from the overall goal to the criteria (C_1, C_2, …, C_n), sub-criteria (C_{11}, C_{12}, …, C_{21}, C_{22}, …, C_{nn}) and alternatives (A_1, A_2, …, A_n). Then, the relative importance of the criteria and alternatives is measured by the expert(s) using the pairwise comparison introduced by Thurston [13] on a scale of 1–9. An important issue that occurs in the pairwise comparisons performing is the inconsistency of the pairwise comparison matrix. It is impossible to ignore the inconsistency in the pairwise comparison matrix [14]. The inconsistency in the decision-making process occurs for various reasons such as lack of decision-makers focus [15]. Performing a large number of pairwise comparisons ($n(n - 1)/2$) to calculate the weights of n criteria is caused to carelessness and fatigue in the decision-makers' minds. This fatigue can lead to inconsistencies in decisions. Therefore, despite the popularity and ease of the AHP, it is sometimes unable to sufficiently control the accuracy of decision-making preferences and the consistency of pairwise comparisons [16]. Pairwise comparison matrix inconsistency is an AHP defect, which may lead to unreasonable ranking results [17].

Rezaei [6] introduced the BWM method requiring fewer pairwise comparison data than the AHP ($2n - 3$) method, while it can make more consistent comparisons. The BWM method allows the decision-maker to produce more reliable results based on previous analyses [18]. In the BWM method, the best (most important) and worst (least important) criteria are determined by decision-makers. Then, the relative importance of other criteria is measured based on the best and worst criteria. Rezaei [6], by comparing the results of BWM and AHP methods, showed that the results of BWM are more accurate than the AHP [6]. However, there are inconsistencies in the BWM method that can affect the decision problems. It is sometimes difficult for decision-makers to identify the best and worst criteria in the first step, because there may be several best or worst criteria of the same importance. To overcome all the mentioned problems of AHP and BWM methods, Haseli et al. [10, 11] introduced the BCM method. The BCM method was able to obtain the weight of criteria full consistency by using fewer pairwise comparisons.

The BCM method similar to the BWM method is vector-based but uses a numerical scale of 1/9 to 9 to determine the relative importance of pairwise comparisons. The BWM method measurement consistencies are based on outputs of pairwise comparisons rather than directly on pairwise comparisons inputs [18]. Unlike previous methods such as AHP, ANP and BWM, the BCM method directly measures inputs of pairwise comparisons to achieve absolute compatibility. By using a framework for measuring pairwise comparison inputs, any inconsistencies in pairwise comparisons can be prevented. In the BCM method, the decision-maker first selected a criterion (preferential) as the base criterion and then performed pairwise comparisons between the base criterion and other criteria. Using the base criterion for comparisons results in better consistency in terms of strength and direction than the other methods. Also, the BCM method is able to complete the pairwise comparison matrix using its own model so that in case of shortage or loss of data, incomplete data can be calculated in full consistency mode by using the BCM framework [10, 11].

2 Base Criterion Method

2.1 Strength and Direction in the Pairwise Comparisons

The main reason for introducing the BCM method is the improvement of the consistency ratio. The weight of the criteria is obtained through pairwise comparisons. In performing pairwise comparisons, the strength and direction of the comparisons are very important [10, 11]. Saaty [7] in the AHP method claims that pairwise comparisons between all criteria are considered independently. To illustrate, he uses the example of football teams playing against each other. His explanation that if there are three criteria for a decision problem (C_1, C_2 and C_3) and the relative importance of C_1 greater than C_2 and the relative importance of C_2 greater than C_3, then there is no requirement for the relative importance of C_1 to be greater than C_3 [7]. On the other hand, the consistency ratio in the AHP method conflicts by performing independent pairwise comparisons. Therefore, this approach to performing the pairwise comparisons leads to inconsistency in pairwise comparison directions [10, 11].

Inconsistency in pairwise comparisons occurs in both statuses of direction and strength. A directional inconsistency occurs when the relative importance is applied inversely to another criterion in pairwise comparisons. For example, if the relative importance of the C_1 is less than C_3 in the above case, it will cause an inconsistency in the direction. Also, inconsistency in the strength occurs when the decision-maker does not accurately evaluate the pairwise comparisons. For example, if the relative importance of C_1 to C_2 is 5 (according to the scale of 1 to 9) and the relative importance of C_2 to C_3 is 1 (equal), then the relative importance of C_1 to C_3 should be 5.

If the decision-maker instead of the value of 5 entered other numbers such as 4 or 6, it will cause inconsistency in strength. Unlike the AHP method, in which inconsistencies in direction and strength are observed, in the BWM method, only inconsistencies in strength occur. In secondary comparisons of the BWM method, directional inconsistency does not occur due to its dependence on the reference comparisons. Inconsistency in strength or direction causes the error and reduced accuracy in determining the optimal weights [10, 11]. Several methods have been introduced over the years to improve the consistency rate [19]. But none of them were able to completely solve the inconsistency problem.

To deal with inconsistencies in strength and direction, it is necessary to reduce the number of pairwise comparisons required and to avoid performing unnecessary pairwise comparisons. Also, a framework is needed to control the inputs of pairwise comparisons instead of evaluating outputs to determine consistency. The BCM method can deal with strength and direction inconsistencies by performing fewer pairwise comparisons required and controlling the inputs of pairwise comparisons. In the BCM method, first base comparisons are performed and if necessary, the final comparisons are calculated based on base comparisons. For more explanation about the base and final comparisons, see Figs. 1 and 2.

Suppose there are six criteria for the MCDM problem. To determine the optimal criteria weights, instead of performing the pairwise comparisons between all criteria, one of the criteria is to select as the base criterion. Then, the relative importance of the base criterion to other criteria is evaluated by using a numerical scale of 1/9 to 9. If necessary, Eq. (1) can be used to complete the matrix of pairwise comparisons.

$$a_{Base,i} \times a_{ij} = a_{Base,j} \tag{1}$$

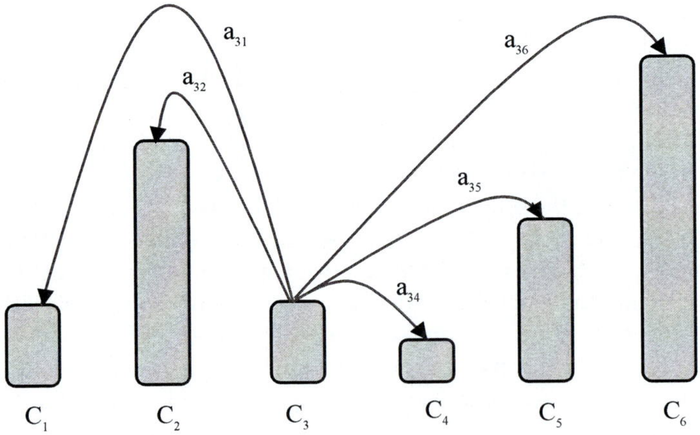

Fig. 1 Base comparisons [10, 11]

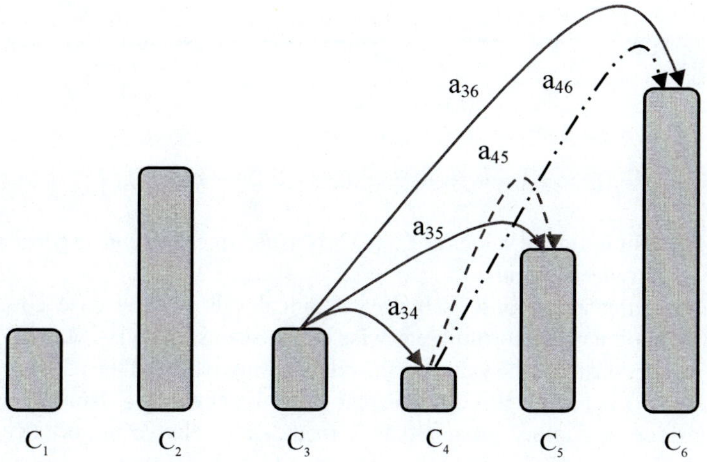

Fig. 2 Final comparisons [10, 11]

The pairwise comparison of a_{ij} indicates the importance of criterion relative to criterion j. Also, the pairwise comparison of a_{ji} indicates the importance of criterion j to criterion i, which is written inversely ($a_{ji} = 1/a_{ij}$). If $a_{ij} > 0$, then criterion i is preferable to criterion j. The pairwise comparison of $a_{Base,i}$ indicates the importance of base criterion relative to criterion i, and the pairwise comparison of $a_{Base,j}$ indicates the importance of the base criterion relative to criterion j.

In fact, the pairwise comparisons of the BCM method are divided into two parts:

1. Base comparisons
2. The final comparisons which calculated using the data of base comparisons.

The pairwise comparison of a_{ij} is a base comparison if i is a base criterion. Otherwise, the pairwise comparison of a_{ij} is one of the final comparisons used to complete the pairwise comparison matrix. The weight values of the criteria can be obtained without performing the final comparisons. The final comparisons are only provided to complete the pairwise comparisons matrix.

As can be seen in Fig. 1, among the six criteria presented, the third criterion (C_3) is selected as the base criterion with the decision-maker preference. Note that there are no restrictions on the choice of base criterion for the decision-maker and the decision-maker by his preference can choose each of the criteria as the base criterion. In the next step, the decision-maker evaluates the relative importance of the base criterion to each of the criteria using a numerical scale of 1/9 to 9.

As shown in Fig. 2, to calculate the relative importance of the pairwise comparison of C_4 to C_5 (dashed line C_{45}), according to Eq. (1), the relative importance of the two base comparisons (C_{34} and C_{35}) is required. For example, if the relative importance value of C_{34} is 2 and the relative importance value of C_{35} is 1/2, then the relative importance of the final comparison of C_{45} will be 1/4. In other words, the value of all final comparisons can be calculated using base comparisons.

$$a_{Base,i} \times a_{ij} = a_{Base,j}$$

$$a_{Base,4} \times a_{45} = a_{Base,5}$$

$$a_{34} \times a_{45} = a_{35} \rightarrow 2 \times a_{45} = 1/2 \rightarrow a_{45} = 1/4$$

The relative importance values of C_4 to C_6 (dashed line C_{46}) can also be calculated like the A_{45} pairwise comparison.

The controlling inputs and assigning a logical value to the base comparisons are important points in performing pairwise comparisons in the BCM method. If the decision-maker considers the value for the relative importance of the base comparison C_3 to C_4 (line a_{34} in Fig. 1), it is obvious that C_3 is higher than C_4. Thus, the decision-maker considers a number greater than 1 to indicate relative importance. On the other hand, C_2, C_5 and C_6 are also larger than C_3. Therefore, the decision-maker must assign a number to a_{34} that does not exceed the numerical scale of 1/9 to 9 in subsequent comparisons. The value of 2 is suitable for the relative importance of a_{34}, but if the decision-maker considered values 3 or 4 for a_{34}, then it will be inconsistent in subsequent comparisons. So if we assign 1/4 to a_{36} (in Fig. 1), then the relative importance of C_4 to C_6 (dashed line a_{46} in Fig. 2) will be 1/12 or 1/16. While the numbers greater than 9 and less than 1/9 are not allowed for assignment to comparisons. To solve this problem, the BCM method considers a framework for controlling the inputs in the base comparisons. According to Eq. (2), the assignment of a number to the relative importance value in the base comparisons must be done in such a way that the values of the final comparisons do not exceed about 1/9 to 9 [10, 11].

$$1/9 \leq \frac{\mathrm{a}Base, \mathrm{j}, \mathrm{j} = 1, 2, \ldots, \mathrm{n}}{\mathrm{aBase}, \mathrm{j}, \mathrm{j} = 1, 2, \ldots, \mathrm{n}} \leq 9 \quad (2)$$

If the values are assigned to the base comparisons according to Eq. (2), the calculated weights will be optimal and fully consistent. Using Eq. (2) to control the inputs of pairwise comparisons, there will be no inconsistencies in direction or strength. The BCM method, instead of evaluating the outputs and trying to improve the inconsistency, produces accurate and reliable results by preventing and controlling the outputs using Eq. (2).

2.2 Steps of BCM

The steps of the BCM method are similar to the BWM method with minor modifications. The BCM method is vector-based and performs only necessary pairwise comparisons to determine the optimal criteria weights. The steps of the BCM method for determining the weight of the criteria include the following four steps [10, 11]:

***Step* (1)** Specify the effective criteria for the decision-making problem.

***Step* (2)** Select one of the criteria as the base criterion.

The base criterion is chosen by the decision-maker and there is no limitation to the decision-maker in choosing it.

***Step* (3)** Perform the base comparisons.

Performing the pairwise comparisons between the base criterion and other criteria using the numerical scale of 1/9 to 9. In this step, for each n criterion n − 1 pairwise comparison is required. The base comparison vector will be in the form of relation (3).

$$a_{Base,j} = (a_{B1}, a_{B2}, a_{B3}, \cdots, a_{Bn}) \tag{3}$$

The principle of Eq. (2) should be considered when performing base comparisons to control inputs. If the relative importance values of any pairwise comparison are inconsistent with the principle of Eq. (2), it will cause inconsistencies in the results. According to the principle of Eq. (2), if all the values for the base comparisons are correct, enter the next step.

***Step* (4)** Determine the optimal weight for each of the criteria.

After performing the base comparisons, Eq. (4) is used to determine the weights of the criteria $(w_1, w_2, w_3, \cdots, w_n)$. To obtain the weight of the criteria, $\frac{w_B}{w_j}$ indicates the pairwise comparison of the base criterion relative to the criterion j, and a_{Bj} indicates the relative importance value of this pairwise comparison.

$$\begin{aligned} &Min\ max\left|\frac{w_B}{w_j} - a_{Bj}\right| \\ &\text{such that,} \\ &\begin{cases} \sum_{j=1}^{n}(w_j) = 1 \\ w_j \geq 0\ for\ all\ j \end{cases} \end{aligned} \tag{4}$$

Considering the deviations leading to the inconsistency, Eq. (4) can be written in the form of Eq. (5).

$$\begin{aligned} &Min\ \xi \\ &\text{such that} \\ &\begin{cases} \left|\frac{w_B}{w_j} - a_{Bj}\right| \leq \xi \\ \sum_{j=1}^{n}(w_j) \\ w_j \geq 0\ for\ all\ j \end{cases} \end{aligned} \tag{5}$$

2.3 BCM Consistency Ratio

Paying attention to the consistency ratio in pairwise comparisons performing is a very important issue. Each of the introduced methods to date has proposed a specific approach for measuring the consistency ratio. These methods (AHP and BWM) measure the consistency ratio through the outputs of comparisons. Therefore, an acceptable value is considered for the inconsistency ratio. This value is often less than 0.1 for the consistency ratio. For example, the BWM method calculates the degree of consistency using Eq. (6). ξ indicates the error rate obtained from the problem-solving algorithm and the consistency index is the rate provided by Rezaei [6].

$$Consistency\ Ratio = \frac{\xi}{Consistency\ Index} \tag{6}$$

The BCM method reduces the error (ξ) resulting from the problem-solving algorithm to zero by controlling the inputs of pairwise comparisons in the base comparisons step. The BCM method uses Eq. (3) to control the inputs of base comparisons. When the value of ξ is zero, the inconsistency ratio will also be zero. The BCM method's goal is to eliminate existing errors to achieve fully consistent weights, which are possible by controlling the inputs of pairwise comparisons. For more information, see solving the examples in Sect. 2.4.

2.4 Examples

This section provides some numerical examples to illustrate the use of the proposed method for decision-making problems. The equations of numerical examples are solved by LINGO software to calculate the optimal weights.

Example 1 A company has considered five criteria for evaluating its staff to selection as a manager. The company intends to assess the importance of each of these criteria in selecting managers. The criteria set include C_1: age, C_2: education, C_3: work experience, C_4: personal competencies and C_5: organizational performance (step 1). The decision-maker needs to select a criterion as the base criterion to evaluate the criteria using the BCM method. The work experience (C3) is selected as the base criterion by the decision-maker (step 2). Then, to determine the relative importance of criteria, pairwise comparisons are performed between the base criterion and other criteria (step 3). The values assigned by the decision-maker to the base comparisons can be seen in Table 1. In this step, the values assigned to the base comparisons are examined according to the principle of Eq. (2). Given that all the values are correct, the optimal weight of the criteria can be calculated.

According to the BCM method mathematical algorithm (Eq. 5), the solution algorithm will be as follows:

Table 1 The base comparisons results

Criteria	C_1	C_2	C_3	C_4	C_5
Base criterion C_3	8	4	1	3	2

$$\begin{aligned}&Min\,\xi\\&such\text{ that}\\&\begin{cases}\left|\frac{w_3}{w_1}-a_{31}\right|\leq\xi\\\left|\frac{w_3}{w_2}-a_{32}\right|\leq\xi\\\left|\frac{w_3}{w_4}-a_{34}\right|\leq\xi\\\left|\frac{w_3}{w_5}-a_{35}\right|\leq\xi\\\sum_{j=1}^{n}(w_j)=1\\w_1\geq0,w_2\geq0,\ w_3\geq0,\ w_4\geq0,\ w_5\geq0\end{cases}\end{aligned}\tag{7}$$

By placing the values of base comparisons in Table 1, the final algorithm will be as follows:

$$\begin{aligned}&Min\,\xi\\&\text{such that}\\&\begin{cases}\left|\frac{w_3}{w_1}-8\right|\leq\xi\\\left|\frac{w_3}{w_2}-4\right|\leq\xi\\\left|\frac{w_3}{w_4}-3\right|\leq\xi\\\left|\frac{w_3}{w_5}-2\right|\leq\xi\\\sum_{j=1}^{n}(w_j)=1\\w_1\geq0,w_2\geq0,\ w_3\geq0,\ w_4\geq0,\ w_5\geq0\end{cases}\end{aligned}\tag{8}$$

By solving the above mathematical Eq. (8) in Lingo software, the weight of the criteria is obtained. The optimal weight of each criterion according to the steps taken is $w_1 = 0.05660377$, $w_2 = 0.1132075$, $w_3 = 0.4528302$, $w_4 = 0.1509434$, $w_5 = 0.2264151$ and $\xi = 0$. As can be seen, the total weight of the criteria is also equal to 1.

Example 2 Suppose seven criteria are effective to product quality. To determine the importance of each criterion, it is necessary to determine the weight of the criteria. To obtain the criteria weights by the BCM method, one of the criteria is chosen as a base criterion (step 2). In this example, the fifth criterion is selected as the base criterion. By performing the base comparisons, values of the relative importance of the base criterion compared to other criteria are obtained (step 3). Table 2 shows the values of the base comparisons.

Table 2 The base comparisons results

Criteria	C_1	C_2	C_3	C_4	C_5	C_6	C_7
Base criterion C_5	2	3	1/3	1/2	1	1/2	1

According to the values of the base comparisons, the mathematical equation of the problem will be as follows:

$$\begin{aligned}
&Min\ \xi \\
&\text{such that} \\
&\begin{cases}
\left|\frac{w_5}{w_1} - 2\right| \le \xi \\
\left|\frac{w_5}{w_2} - 3\right| \le \xi \\
\left|\frac{w_5}{w_3} - 1/3\right| \le \xi \\
\left|\frac{w_5}{w_4} - 1/2\right| \le \xi \\
\left|\frac{w_5}{w_6} - 1/3\right| \le \xi \\
\left|\frac{w_5}{w_7} - 1\right| \le \xi \\
\sum_{j=1}^{n} (w_j) \\
w_1 \ge 0,\ w_2 \ge 0,\ w_3 \ge 0,\ w_4 \ge 0, \\
w_5 \ge 0,\ w_6 \ge 0,\ w_7 \ge 0
\end{cases}
\end{aligned} \tag{9}$$

By solving the mathematical Eq. (9) in Lingo software, the results will be $w_1 = 0.05084746$, $w_2 = 0.3389831$, $w_3 = 0.3050847$, $w_4 = 0.2033898$, $w_5 = 0.1016949$, $w_6 = 0.2033898$, $w_5 = 0.1016949$ and $\xi = 0$.

Note 1 The important point in this example is the existence of several criteria with equal importance. Sometimes there may be several criteria of equal importance in the decision-making process. The equality of the importance of several criteria should not influence the decision-maker in choosing the base criterion.

Note 2 By comparing the number of pairwise comparisons required to obtain the optimal weights for this problem, the advantage of the BCM method is able to realize so that for seven criteria by using the AHP method 21 pairwise comparisons, by using the BWM method 11 pairwise comparisons and by using the BCM method only 6 pairwise comparisons are required.

Example 3 This example provided a mode to compare the answers and analyze the results. The totality of the example and the values of pairwise comparisons for solving this problem are derived from the numerical examples of [6] and Haseli et al. [10, 11]. A company has identified three criteria for choosing the optimal mode of transportation of products to the market. The decision-maker wants to determine the

Table 3 The base comparisons results

Criteria	C_1	C_2	C_3
Base criterion C3	8	2	1

impact (weight) of each criterion on the optimal model. The criteria identified are as follows:

C_1 Load flexibility
C_2 Accessibility
C_3 Cost

The cost criterion is chosen as the base criterion by the decision-maker. Table 3 shows the base comparisons made by the decision-maker.

According to the BCM method algorithm (Eq. 5), the solution algorithm will be as follows:

$$\begin{aligned}&Min\ \xi\\&\text{such that}\\&\begin{cases}\left|\frac{w_3}{w_1}-a_{31}\right|\leq\xi\\\left|\frac{w_3}{w_2}-a_{32}\right|\leq\xi\\\sum_{j=1}^{n}(w_j)=1\\w_1\geq 0,\ w_2\geq 0,\ w_3\geq 0\end{cases}\end{aligned}\tag{10}$$

By replacing the values of base comparisons in Eq. (10), the final algorithm will be as follows:

$$\begin{aligned}&Min\ \xi\\&such\ that\\&\begin{cases}\left|\frac{w_3}{w_1}-8\right|\leq\xi\\\left|\frac{w_3}{w_2}-2\right|\leq\xi\\w_1+w_2+w_3=1\\w_1\geq 0,\ w_2\geq 0,\ w_3\geq 0\end{cases}\end{aligned}\tag{11}$$

The optimal weight of each criterion according to the evaluation of the decision-maker is $w_1 = 0.07692308$, $w_2 = 0.3076923$ and $w_3 = 0.6153846$. Also, the error rate is zero ($\xi = 0$), which indicates full consistency. According to Eq. (2), the complete pairwise comparison matrix can be calculated (can be seen in Table 4).

Note 3 If the weights obtained from the BCM method are compared with the result weights of the BWM method, we will notice a difference in the calculated values. The main reason for this difference in the value of obtained weights is the error rate of 0.26 that occurred in the BWM pairwise comparisons. Table 5 shows the obtained results of solved examples by BWM [6] and BCM [10, 11] methods

Table 4 The values of the complete pairwise comparison matrix

Criteria	Load flexibility	Accessibility	Cost
Load flexibility	1	1/4	1/8
Accessibility	4	1	1/2
Cost	8	2	1

Table 5 The results of BCM and BWM

Weights	BCM	BWM
Load flexibility	0.07692308	0.0714
Accessibility	0.3076923	0.03387
Cost	0.6153846	0.5899
ξ	0.00	0.26

Note 4 In Table 6, all the values for s31 and a32 are considered so that the different weights of the criteria and ξ for the consistency ratio can be seen. Table 6 shows that the weight of the criteria changes as the value of each of the base comparisons changes. It is observed that in some cases, the values for a13 and a23 in Table 6 are not consistent with Eq. (2), but the error (ξ) is zero. In fact, Eq. (2) is used to control inputs, ensure consistency and obtain fully optimal weights

In real-life MCDM problems, ambiguity in evaluating comparisons is often due to a lack of complete knowledge, and ambiguity of decision-makers in qualitative judgment is quite common, so sometimes using clear values to evaluate criteria is insufficient [20]. Thus, based on the uncertainty and complexity of the goal and human thinking, fuzzy recruitment variables may reflect a better approach to addressing the MCDM's scientific topics [2]. Fuzzy set theory by [21] helps decision-maker to deal with vague, inaccurate and subjective data that characterize human behavior and judgment [22].

Haseli et al. [10, 11] proposed a fuzzy BCM method based on triangular fuzzy numbers for ambiguous decision problems that have subjective criteria. The fuzzy BCM method is similar to the BCM method and is able to obtain the optimal weight for the subjective criteria. See (Haseli et al. 2020) for more details and familiarity.

3 Conclusion

The BCM method uses pairwise comparisons to obtain the weight of the criteria. The results show that the BCM method is more reliable than other MCDM methods and provides more accurate results. Also, the BCM method needed to perform fewer pairwise comparisons to obtain the weight of the criteria (n − 1) than the other MCDM methods. An important principle that makes the BCM method superior to other existing MCDM methods is the control of inputs of pairwise comparisons instead of checking outputs. In the BCM method, the inputs of pairwise comparisons

Table 6 All values of w_1, w_2, w_3 and ξ for each all j in base comparisons

a_{31}		a_{32}								
		1	2	3	4	5	6	7	8	9
1	ξ	1.4×10^{-6}	0.0	0.0	1.6×10^{-6}	2.1×10^{-6}	1.6×10^{-6}	0.0	2×10^{-6}	1.7×10^{-6}
	w_1	0.333	0.4	0.429	0.444	0.455	0.462	0.467	0.471	0.474
	w_2	0.333	0.2	0.143	0.111	0.091	0.077	0.067	0.059	0.053
	w_3	0.333	0.4	0.429	0.444	0.455	0.462	0.467	0.471	0.474
2	ξ	0.0	0.0	0.0	0.0	0.0	0.0	0.0	0.0	0.0
	w_1	0.2	0.25	0.273	0.286	0.294	0.3	0.304	0.308	0.310
	w_2	0.4	0.25	0.182	0.143	0.118	0.1	0.087	0.077	0.069
	w_3	0.4	0.5	0.545	0.571	0.588	0.6	0.609	0.615	0.621
3	ξ	0.0	0.0	0.0	0.0	0.0	0.0	0.0	0.0	0.0
	w_1	0.143	0.182	0.2	0.211	0.217	0.222	0.226	0.229	0.231
	w_2	0.429	0.273	0.2	0.158	0.13	0.111	0.097	0.086	0.077
	w_3	0.429	0.545	0.6	0.632	0.652	0.666	0.677	0.686	0.692
4	ξ	1.6×10^{-6}	0.0	0.0	0.0	0.0	0.0	0.0	0.0	0.0
	w_1	0.111	0.143	0.158	0.166	0.172	0.176	0.179	0.182	0.184
	w_2	0.444	0.286	0.211	0.167	0.138	0.118	0.103	0.091	0.082
	w_3	0.444	0.571	0.632	0.667	0.69	0.706	0.718	0.727	0.735
5	ξ	0.0	0.0	0.0	0.0	1.3×10^{-6}	6.5×10^{-6}	0.0	0.0	1.3×10^{-6}
	w_1	0.091	0.118	0.13	0.138	0.143	0.146	0.149	0.151	0.153
	w_2	0.455	0.294	0.217	0.172	0.143	0.122	0.106	0.094	0.085
	w_3	0.455	0.588	0.652	0.69	0.714	0.732	0.745	0.755	0.763

(continued)

Table 6 (continued)

a_{31}		a_{32}								
		1	2	3	4	5	6	7	8	9
6	ξ	1.3×10^{-6}	0.0	0.0	0.0	1.3×10^{-6}	1.3×10^{-6}	0.0	0.0	0.0
	w_1	0.077	0.1	0.111	0.118	0.122	0.125	0.127	0.129	0.13
	w_2	0.462	0.3	0 .222	0.176	0.146	0.125	0.109	0.097	0.087
	w_3	0.462	0.6	0.666	0.706	0.732	0.75	0.764	0.774	0.783
7	ξ	0.0	0.0	0.0	0.0	0.0	1.3×10^{-6}	0.0	0.0	0.0
	w_1	0.067	0.087	0.097	0.103	0.106	0.109	0.111	0.113	0.114
	w_2	0.467	0.304	0.226	0.179	0.149	0.127	0.111	0.099	0.089
	w_3	0.467	0.609	0.677	0.718	0.745	0.764	0.778	0.789	0.797
8	ξ	0.0	0.0	0.0	0.0	0.0	0.0	0.0	1.3×10^{-6}	1.3×10^{-6}
	w_1	0.059	0.077	0.086	0.091	0.094	0.097	0.099	0.1	0.101
	w_2	0.471	0.308	0.229	0.182	0.151	0.129	0.113	0.1	0.09
	w_3	0.471	0.615	0.686	0.727	0.755	0.774	0.789	0.8	0.809
9	ξ	0.0	0.0	0.0	0.0	0.0	1.3×10^{-6}	0.0	0.0	1.3×10^{-6}
	w_1	0.053	0.069	0.077	0.082	0.085	0.087	0.089	0.09	0.091
	w_2	0.474	0.310	0.231	0.184	0.153	0.13	0.114	0.101	0.091
	w_3	0.474	0.621	0.692	0.735	0.763	0.783	0.797	0.809	0.818
		1	1/2	1/3	1/4	1/5	1/6	1/7	1/8	1/9
1	ξ	1.4×10^{-6}	0.0	0.0	6.9×10^{-7}	0.0	1.7×10^{-7}	0.0	0.0	0.0
	w_1	0.333	0.25	0.2	0.167	0.143	0.125	0.111	0.1	0.091

(continued)

Table 6 (continued)

a_{31}		a_{32}								
		1	2	3	4	5	6	7	8	9
	w_2	0.333	0.5	0.6	0.667	0.714	0.75	0.111	0.8	0.818
	w_3	0.333	0.25	0.2	0.167	0.143	0.125	0.778	0.1	0.091
2	ξ	0.0	0.0	0.0	2.8×10^{-7}	1.5×10^{-7}	0.0	8×10^{-7}	0.0	0.0
	w_1	0.2	0.143	0.111	0.091	0.077	0.067	0.059	0.053	0.048
	w_2	0.4	0.571	0.667	0.727	0.769	0.8	0.824	0.842	0.857
	w_3	0.4	0.286	0.222	0.182	0.154	0.133	0.118	0.105	0.095
3	ξ	0.0	0.0	0.0	9.3×10^{-7}	7.6×10^{-7}	2.1×10^{-7}	3×10^{-7}	1.7×10^{-7}	4×10^{-7}
	w_1	0.143	0.1	0.077	0.062	0.053	0.045	0.04	0.036	0.032
	w_2	0.429	0.6	0.692	0.750	0.789	0.818	0.84	0.857	0.871
	w_3	0.429	0.3	0 .231	0.188	0.158	0.136	0.12	0.107	0.097
4	ξ	1.6×10^{-6}	2.7×10^{-7}	0.0	6.1×10^{-7}	5.1×10^{-7}	7.5×10^{-7}	2.6×10^{-7}	4.3×10^{-7}	9.4×10^{-7}
	w_1	0.111	0.077	0.059	0.048	0.8	0.034	0.03	0.027	0.024
	w_2	0.444	0.615	0.706	0.762	0.04	0.828	0.848	0.865	0.878
	w_3	0.444	0.308	0.235	0.19	0.16	0.138	0.121	0.108	0.098
5	ξ	2.1×10^{-6}	4.9×10^{-7}	0.0	4.6×10^{-7}	3.9×10^{-7}	5.9×10^{-7}	2.2×10^{-7}	4×10^{-7}	8.8×10^{-7}
	w_1	0.091	0.062	0.048	0.039	0.323	0.139	0.024	0.022	0.02
	w_2	0.455	0.625	0.714	0.769	0.806	0.833	0.854	0.87	0.882
	w_3	0.455	0.313	0.238	0.192	0.161	0.139	0.122	0.109	0.098
6	ξ	1.6×10^{-6}	0.0	1.3×10^{-6}	0.0	6.5×10^{-7}	1.1×10^{-6}	0.0	0.0	9.7×10^{-7}

(continued)

Table 6 (continued)

a_{31}		a_{32}								
		1	2	3	4	5	6	7	8	9
	w_1	0.077	0.053	0.04	0.032	0.027	0.023	0.021	0.018	0.016
	w_2	0.462	0.632	0.72	0.774	0.811	0.837	0.857	0.873	0.885
	w_3	0.462	0.316	0.24	0.194	0.162	0.14	0.122	0.109	0.098
7	ξ	0.0	1.5×10^{-6}	9.2^{-7}	0.0	0.0	7.7^{-7}	0.0	1×10^{-6}	0.0
	w_1	0.067	0.045	0.035	0.028	0.023	0.02	0.018	0.016	0.014
	w_2	0.467	0.636	0.724	0.778	0.814	0.84	0.86	0.875	0.887
	w_3	0.467	0.318	0.241	0.194	0.163	0.14	0.123	0.109	0.099
8	ξ	2×10^{-6}	0.0	7.4×10^{-7}	0.0	0.0	6×10^{-7}	0.0	7.8^{-7}	0.0
	w_1	0.059	0.04	0.03	0.024	0.02	0.018	0.015	0.014	0.012
	w_2	0.471	0.64	0.727	0.781	0.816	0.842	0.862	0.877	0.889
	w_3	0.471	0.32	0.242	0.195	0.163	0.14	0.123	0.11	0.099
9	ξ	1.7×10^{-6}	0.0	6.4×10^{-7}	0.0	0.0	0.0	0.0	6.4×10^{-7}	0.0
	w_1	0.053	0.036	0.027	0.022	0.018	0.016	0.014	0.012	0.011
	w_2	0.474	0.643	0.73	0.783	0.818	0.844	0.863	0.878	0.89
	w_3	0.474	0.321	0.243	0.196	0.164	0.14	0.123	0.11	0.099
		1	2	3	4	5	6	7	8	9
1	ξ	1.4×10^{-6}	0.0	0.0	1.6×10^{-6}	2.1×10^{-6}	1.6×10^{-6}	0.0	2×10^{-6}	1.7×10^{-6}
	w_1	0.333	0.4	0.429	0.444	0.455	0.462	0.467	0.471	0.474
	w_2	0.333	0.2	0.143	0.111	0.091	0.077	0.067	0.059	0.053

(continued)

Table 6 (continued)

a_{31}		a_{32}								
		1	2	3	4	5	6	7	8	9
	w_3	0.333	0.4	0.429	0.444	0.455	0.462	0.467	0.471	0.474
1/2	ξ	8.2×10^{-7}	0.0	0.0	1.2×10^{-6}	1.6×10^{-6}	0.0	0.0	9.8×10^{-7}	8.7×10^{-7}
	w_1	0.5	0.571	0.6	0.615	0.625	0.632	0.636	0.64	0.643
	w_2	0.25	0.163	0.1	0.077	0.313	0.053	0.046	0.04	0.036
	w_3	0.25	0.286	0.3	0.308	0.062	0.316	0.318	0.32	0.321
1/3	ξ	1.4×10^{-6}	1×10^{-6}	0.0	0.0	0.0	1.6×10^{-9}	0.0	0.0	0.0
	w_1	0.6	0.667	0.692	0.706	0.714	0.72	0.724	0.727	0.73
	w_2	0.2	0.111	0.077	0.059	0.048	0.04	0.034	0.03	0.027
	w_3	0.2	0.222	0.231	0.235	0.238	0.24	0.242	0.243	0.243
1/4	ξ	0.0	8.7×10^{-7}	6.7×10^{-7}	7×10^{-7}	0.0	3.2×10^{-7}	0.0	0.0	0.0
	w_1	0.667	0.727	0.75	0.762	0.769	0.774	0.778	0.781	0.783
	w_2	0.167	0.091	0.063	0.048	0.038	0.032	0.028	0.024	0.022
	w_3	0.166	0.182	0.187	0.19	0.192	0.194	0.194	0.195	0.196
1/5	ξ	0.0	1.2×10^{-6}	0.0	0.0	0.0	0.0	0.0	0.0	8.7×10^{-7}
	w_1	0.714	0.625	0.556	0.5	0.455	0.417	0.385	0.357	0.333
	w_2	0.143	0.25	0.333	0.4	0.455	0.5	0.538	0.571	0.6
	w_3	0.143	0.125	0.111	0.1	0.09	0.083	0.077	0.072	0.067
1/6	ξ	0.0	0.0	1.2×10^{-6}	0.0	0.0	0.0	0.0	0.0	0.0
	w_1	0.75	0.667	0.6	0.545	0.5	0.462	0.429	0.4	0.375

(continued)

Table 6 (continued)

a_{31}		a_{32}								
		1	2	3	4	5	6	7	8	9
	w_2	0.125	0.222	0.3	0.364	0.417	0.462	0.5	0.533	0.563
	w_3	0.125	0.111	0.1	0.091	0.083	0.077	0.071	0.067	0.062
1/7	ξ	0.0	0.0	0.0	0.0	0.0	0.0	0.0	0.0	0.0
	w_1	0.778	0.7	0.636	0 .583	0.538	0.5	0.467	0.437	0.412
	w_2	0.111	0.2	0.273	0.333	0.385	0.429	0.467	0.5	0.53
	w_3	0.111	0.1	0.091	0.083	0.077	0.071	0.666	0.063	0.058
1/8	ξ	0.0	0.0	0.0	0.0	0.0	1.5×10^{-6}	0.0	0.0	0.0
	w_1	0.8	0.727	0.667	0.615	0.571	0.533	0.5	0.471	0.444
	w_2	0.1	0.182	0.25	0.308	0.357	0.4	0.437	0.471	0.5
	w_3	0.1	0.091	0.083	0.077	0.072	0.067	0.063	0.059	0.556
1/9	ξ	0.0	0.0	0.0	0.0	1×10^{-6}	0.0	1.2×10^{-6}	0.0	0.0
	w_1	0.818	0.75	0.692	0.643	0.6	0.562	0.529	0.5	0.474
	w_2	0.091	0.167	0.231	0.286	0.333	0.375	0.412	0.444	0.474
	w_3	0.091	0.083	0.077	0.071	0.067	0.063	0.059	0.056	0.052
		1	1/2	1/3	1/4	1/5	1/6	1/7	1/8	1/9
1	ξ	1.4×10^{-6}	0.0	0.0	6.9×10^{-7}	0.0	1.7×10^{-7}	0.0	0.0	0.0
	w_1	0.333	0.25	0.2	0.167	0.143	0.125	0.111	0.1	0.091
	w_2	0.333	0.5	0.6	0.667	0.714	0.75	0.111	0.8	0.818

(continued)

Table 6 (continued)

a_{31}		a_{32}								
		1	2	3	4	5	6	7	8	9
	w_3	0.333	0.25	0.2	0.167	0.143	0.125	0.778	0.1	0.091
1/2	ξ	8.2×10^{-7}	0.0	1.9×10^{-6}	0.0	1.4×10^{-6}	1.4×10^{-6}	0.0	0.0	4.3×10^{-6}
	w_1	0.5	0.4	0.333	0.286	0.25	0.222	0.2	0.182	0.167
	w_2	0.25	0.4	0.5	0.571	0.625	0.667	0.7	0.727	0.75
	w_3	0.25	0.2	0.167	0.143	0.125	0.111	0.1	0.091	0.083
1/3	ξ	1.4×10^{-6}	1×10^{-6}	2.4×10^{-7}	0.0	1.8×10^{-6}	0.0	0.0	0.0	0.0
	w_1	0.6	0.5	0.429	0.375	0.333	0.3	0.273	0.25	0.231
	w_2	0.2	0.333	0.429	0.5	0.556	0.6	0.636	0.667	0.692
	w_3	0.2	0.167	0.143	0.125	0.111	0.1	0.091	0.083	0.077
1/4	ξ	0.0	0.0	0.0	1.5×10^{-7}	0.0	0.0	0.0	5.3×10^{-7}	0.0
	w_1	0.667	0.571	0.5	0.444	0 .4	0.364	0.333	0.308	0.286
	w_2	0.167	0.286	0.375	0.445	0.5	0.545	0.583	0.615	0.643
	w_3	0.167	0.143	0.125	0.111	0.1	0.091	0.084	0.077	0.071
1/5	ξ	0.0	1.2×10^{-6}	0.0	0.0	0.0	0.0	0.0	0.0	8.7×10^{-7}
	w_1	0.714	0.625	0.556	0.5	0.545	0.417	0.385	0.357	0.333
	w_2	0.143	0.25	0.333	0.2	0.545	0.5	0.538	0.571	0.6
	w_3	0.143	0.125	0.111	0.1	0.091	0.083	0.077	0.072	0.067
1/6	ξ	0.0	0.0	1.2×10^{-6}	0.0	0.0	0.0	0.0	0.0	0.0
	w_1	0.75	0.667	0.6	0.545	0.5	0.462	0.429	0.4	0.375

(continued)

Table 6 (continued)

a_{31}		a_{32}								
		1	2	3	4	5	6	7	8	9
	w_2	0.125	0.222	0.3	0.364	0.417	0.461	0.5	0.533	0.563
	w_3	0.125	0.111	0.1	0.091	0.083	0.077	0.071	0.067	0.062
1/7	ξ	0.0	0.0	0.0	0.0	0.0	0.0	0.0	0.0	0.0
	w_1	0.778	0.7	0.636	0.583	0.538	0.5	0.467	0.438	0.412
	w_2	0.111	0.2	0.273	0.333	0.385	0.429	0.467	0.5	0.529
	w_3	0.111	0.1	0.091	0.083	0.077	0.071	0.066	0.062	0.059
1/8	ξ	0.0	0.0	0.0	0.0	0.0	1.5×10^{-6}	0.0	0.0	0.0
	w_1	0.8	0.727	0.667	0.615	0.572	0.533	0.5	0.471	0.444
	w_2	0.1	0.182	0.25	0.308	0.357	0.4	0.437	0.471	0.5
	w_3	0.1	0.091	0.073	0.077	0.071	0.067	0.63	0.059	0.056
1/9	ξ	0.0	0.0	0.0	0.0	1×10^{-6}	0.0	1.2×10^{-6}	0.0	0.0
	w_1	0.818	0.75	0.692	0.643	0.6	0.562	0.529	0.5	0.474
	w_2	0.091	0.167	0.231	0.286	0.333	0.375	0.412	0.444	0.474
	w_3	0.091	0.083	0.077	0.071	0.067	0.063	0.059	0.056	0.053

are controlled using Eq. (2) so that inconsistencies in direction and strength do not occur. Due to the full consistency of the results based on input controlling, the weights obtained using the BCM method will be quite accurate. This method is simple and requires fewer equations to obtain the weight of the criteria. Using the BCM method, decision-makers will have the option of selecting each of the criteria as the base criterion for performing pairwise comparisons. Using the BCM method, the values of missing data in the incomplete pairwise comparison matrix can be calculated under full compatibility conditions.

One of the limitations of this method is reduced attractiveness when there are a large number of criteria. It is recommended that when dealing with decision problems that affect more than 12 criteria, the criteria be divided into different groups or criteria of lesser importance be removed. Also, sometimes the inputs of pairwise comparisons may not be based on Eq. (2) but still, the results are consistent (as can be seen in Table 6). It is recommended that Eq. (2) is always considered by the decision-makers as the basic element for controlling inputs of pairwise comparisons.

References

1. Zavadskas, E.K., Antucheviciene, J., Hajiagha, S.H., Hashemi, S.S.: Extension of weighted aggregated sum product assessment with interval-valued intuitionistic fuzzy numbers (WASPAS-IVIF). Appl. Soft Comput. **24**, 1013–1021 (2014)
2. Guo, S., Zhao, H.: Fuzzy best-worst multi-criteria decision-making method and its applications. Knowl.-Based Syst. **121**, 23–31 (2017)
3. Kahraman, C., Onar, S.C., Oztaysi, B.: Fuzzy multicriteria decision-making: a literature review. Int. J. Comput. Intell. Syst. **8**(4), 637–666 (2015)
4. Mardani, A., Nilashi, M., Zavadskas, E.K., Awang, S.R., Zare, H., Jamal, N.M.: Decision making methods based on fuzzy aggregation operators: Three decades review from 1986 to 2017. Int. J. Inf. Technol. Decis. Mak. **17**(02), 391–466 (2018)
5. Liao, H., Wu, X., Liang, X., Xu, J., Herrera, F.: A new hesitant fuzzy linguistic ORESTE method for hybrid multicriteria decision making. IEEE Trans. Fuzzy Syst. **26**(6), 3793–3807 (2018)
6. Rezaei, J.: Best-worst multi-criteria decision-making method. Omega **53**, 49–57 (2015)
7. Saaty, T.L.: A scaling method for priorities in hierarchical structures. J. Math. Psychol. **15**(3), 234–281 (1977)
8. Saaty, T.L.: Theory and applications of the analytic network process: decision making with benefits, opportunities, costs, and risks. RWS publications (2005)
9. Rezaei, J.: Best-worst multi-criteria decision-making method: some properties and a linear model. Omega **64**, 126–130 (2016)
10. Haseli, G., Sheikh, R., Sana, S.S.: Base-criterion on multi-criteria decision-making method and its applications. Int. J. Manage. Sci. Eng. Manage. **15**(2), 79–88 (2020)
11. Haseli, G., Sheikh, R., Sana, S.S.: Extension of base-criterion method based on fuzzy set theory. Int. J. Appl. Comput. Math. **6**, 54 (2020)
12. Saaty, T.L.: What is the Analytic Hierarchy Process? In: Mathematical Models for ,rt, pp. 109–121. Springer, Berlin, Heidelberg (1988)
13. Thurstone, L.L.: A law of comparative judgment. Psychol. Rev. **34**(4), 273 (1927)
14. Herman, M.W., Koczkodaj, W.W.: A Monte Carlo study of parwise comparison. Inf. Process. Lett. **57**(1), 25–29 (1996)

15. Forman, E.H., Selly, M.A.: Decision by Objectives: How to Convince Others that you are Right. World Scientific (2001)
16. Ishizaka, A., Nguyen, N.H.: Calibrated fuzzy AHP for current bank account selection. Expert Syst. Appl. **40**(9), 3775–3783 (2013)
17. Dong, J., Wan, S., Chen, S.M.: Fuzzy best-worst method based on triangular fuzzy numbers for multi-criteria decision-making. Inf. Sci. **547**, 1080–1104 (2020)
18. Liang, F., Brunelli, M., Rezaei, J.: Consistency issues in the best worst method: measurements and thresholds. Omega **96**, 102175 (2020)
19. Harker, P.T.: Alternative modes of questioning in the analytic hierarchy process. Math. Model. **9**(3–5), 353–360 (1987)
20. Fei, L., Lu, J., Feng, Y.: An extended best-worst multi-criteria decision-making method by belief functions and its applications in hospital service evaluation. Comput. Ind. Eng. **142**, 106355 (2020)
21. Zadeh, L.A.: Fuzzy sets. In: Fuzzy Sets, Fuzzy Logic, and Fuzzy Systems, pp. 394–432 (1965)
22. Rashidi, K., Cullinane, K.: A comparison of fuzzy DEA and fuzzy TOPSIS in sustainable supplier selection: Implications for sourcing strategy. Expert Syst. Appl. **121**, 266–281 (2019)
23. Rezaei, J.: Piecewise linear value functions for multi-criteria decision-making. Expert Syst. Appl. **98**, 43–56 (2018)
24. Saaty, T.L.: How to make a decision: the analytic hierarchy process. Eur. J. Oper. Res. **48**(1), 9–26 (1990)

DEX (Decision EXpert): A Qualitative Hierarchical Multi-criteria Method

Marko Bohanec

Abstract DEX (Decision EXpert) is a hierarchical, qualitative, rule-based, multi-criteria decision modeling method. It combines multi criteria decision analysis with artificial intelligence and is particularly suited for sorting/classification decision problems. DEX puts special attention on the transparency, comprehensibility, consistency, and completeness of decision models, as well as on methods for the analysis, justification, and explanation of decisions. The approach relies on using software tools that actively support the decision maker in both the creation and utilization stages of the process. Since its inception in the 1980s, DEX has been successfully applied in hundreds of real-world decision projects in various areas, including economy, ecology, agronomy, medicine, and health care. In the last decade, there is an increasing trend of including DEX models in decision support systems. In this chapter, DEX is described from the theoretical and practical viewpoint and further explained in terms of motivation, history, software, applications, and method extensions. The presentation is supported by three examples: a didactic example of employee selection and two real-world industrial applications of choosing a raw-material location and assessing electric energy production technologies, respectively.

Keywords Decision Expert (DEX) · Multi-criteria modeling · Qualitative decision model · Decision rules · Decision support system · DEXi

List of Symbols

$\mathcal{A} = \{A_1, A_2, \ldots, A_q\}$ Alternatives

M. Bohanec (✉)
Department of Knowledge Technologies, Jožef Stefan Institute, Ljubljana, Slovenia
e-mail: marko.bohanec@ijs.si

University of Nova Gorica, Nova Gorica, Slovenia

A. J. Kulkarni (ed.), *Multiple Criteria Decision Making*, Studies in Systems, Decision and Control 407, https://doi.org/10.1007/978-981-16-7414-3_3

$A_i \in \mathcal{A}, A_i = \{a_{x,i} \in E_x, \forall x \in X\}$	An alternative
$a_{x,i} \in A_i$	Value of A_i assigned to attribute x
$B_x \subset D_x$	Subset of bad values of attribute x
$C_y = \prod_{x \in S(y)} D_x$	Domain of f_y
D_x	Value scale of attribute x
E_x	Range of values that can be assigned to attribute x
E_y	Range of f_y
$e = (\mathbf{x}, y) \in T_y$	Elementary decision rule, an entry in T_y
F	Set of aggregation functions of a DEX model
$\mathcal{F}_y$	Fuzzy distributions over D_y
$f_y : C_y \rightarrow E_y$	Aggregation function associated with attribute y
g_y	An approximation of f_y
$G_x \subset D_x$	Subset of good values of attribute x
$I_i \subset A_i$	Subset of values of A_i, assigned to input attributes
I_y	Set of intervals over D_y
$m_x = \|D_x\|$	Number of categories of scale D_x
$M = (X, D, S, F)$	A DEX model
$N_x \subset D_x$	Subset of neutral values of attribute x
$O_i \subset A_i$	Subset of values of A_i, assigned to output (aggregate) attributes
$\mathrm{ord}(v_{x,i}) = i$	Ordinal value of $v_{x,i}$
$P(x)$	Set of parents of attribute x
$\mathcal{P}_y$	Probability distributions over D_y
$r_y = \|C_y\|$	Size of C_y and the corresponding T_y
$S : X \rightarrow 2^x$	Descendant function
$S(x)$	Set of descendants of attribute x
$\mathcal{S}_y$	The power set of D_y
$T_y = \{(\mathbf{x}, y), \mathbf{x} \in C_y, y \in E_y\}$	Decision table associated with attribute y
$v_{x,i} \in D_x$	i-th qualitative value (category) of attribute x
$v_{x,i} \preccurlyeq v_{x,j}$	Weak preference relation
$w, w_i \in \mathcal{R}$	Relative weight (importance) of an attribute
$x, x_i, y \in X$	An attribute
$X = \{x_1, x_2, \ldots, x_n\}$	Set of attributes
$\omega \in [-0.5, +0.5]$	An offset to qualitative value v

Abbreviations

3D	Three-dimensional
AHP	Analytic Hierarchy Process, an MCDM method

AQ	Algorithm Quasi-optimal, a machine rule learning algorithm
CDPC	Consistency-Driven Pairwise Comparisons, an MCDM method
DECMAK	DECision MAKing, an early predecessor of DEX
DEX	Decision EXpert, a qualitative MCDM method
DEXi	Software implementing the DEX method
DRSA	Dominance-based Rough Set Analysis, an MCDM approach
DSS	Decision Support System
ELECTRE	ELimination Et Choix Traduisant la REalité (ELimination Et Choice Translating REality), a family of MCDM methods
HINT	Hierarchical INduction Tool, a machine-learning method for developing DEX models from data
MACBETH	Measuring Attractiveness by a Categorical Based Evaluation Technique, an MCDM method
MCDM	Multi-Criteria Decision Modeling
MCHP	Multi-Criteria Hierarchy Process, a hierarchical MCDM approach
QQ	Qualitative-Quantitative, an approach to ranking of alternatives using a DEX model

1 Introduction

DEX (Decision EXpert) [19] is a multi-criteria decision modeling (MCDM) method, conceived in the 1980s as a fusion of multi-criteria decision analysis and artificial intelligence. From MCDM, it adopted the ideas of modeling decision situations using multiple criteria, structuring and decomposing complex decision problems in smaller and less complex sub-problems, and solving problems through evaluation and analysis of decision alternatives. From artificial intelligence, it primarily adopted concepts used in expert systems: using qualitative (symbolic) variables, representing decision knowledge in terms of "if–then" rules, handling imprecision and uncertainty, emphasizing the transparency of decision models, and facilitating the explanation of results. DEX also includes some elements of machine learning, e.g., for constructing compact decision rules from decision tables.

According to the classification in [49], DEX belongs to the category of *full aggregation* or "American school" methods. This approach is characterized by using an explicit multi-criteria model, which is developed first, more or less independently from individual decision alternatives. These alternatives are then evaluated by the model, first by scoring them for each criterion and then aggregating these evaluations into a global score.

DEX is also characterized as follows [22]:

1. DEX is *hierarchical*: a DEX model consists of hierarchically structured *attributes* (in MCDM, also called *criteria* or *performance variables*). In this aspect, DEX is similar to other hierarchical MCDM methods [3, 45], such as AHP [85] and MCHP [34].
2. DEX is *qualitative*: all attributes in a DEX model are symbolic, taking values that are words rather than numbers, such as "bad", "medium", "excellent", "low", or "high". This relates DEX to verbal decision analysis [65], linguistic MCDM [31, 42], and MCDM methods that use words, such as MACBETH [2].
3. DEX is *rule-based*: hierarchical aggregation of values is defined with decision rules, acquired and represented in the form of decision tables. In this way, DEX is most similar to Dominance-Based Rough Set Analysis [43], which also uses decision tables and constructs decision rules from them.

Given its qualitative nature, DEX is particularly suitable for *sorting* [82] or *classification* [39, 53] decision tasks, which are aimed at assigning each decision alternative to the one category among a family of predefined categories. These categories can be preferentially ordered (sorting) or not (classification). There are also variations of DEX adapted for the ranking problem [8, 60].

In the remaining part of this chapter, the DEX method is presented in detail. After a brief historical overview, the concept of a DEX model is formally defined and illustrated using an employee selection example. This is followed by dynamic aspects of DEX, which are reflected in algorithms that support the creation and modification of decision tables and perform the evaluation and analysis of alternatives. Practical applications of DEX are reviewed and illustrated by two real-world industrial examples: choosing a clay-pit location and assessing electric energy production technologies. Final sections include notes on DEX extensions and a summary.

2 DEX Method and Software: A Brief History

The development of DEX can be traced back to Efstathiou and Rajkovič [40] who proposed using fuzzy sets [93, 94] and fuzzy inference rules to represent and evaluate decision alternatives. The authors also suggested representing decision knowledge in terms of a decision table together with fuzzy operators. The following development of DEX was mainly continued at the Jožef Stefan Institute, Ljubljana, Slovenia, where elements of expert systems [50, 73] and machine learning [30, 64] were gradually added to the basic concepts, leaving the fuzzy aspects somewhat aside. The method, presented by presented by Rajkovič, et al. [77] and Bohanec and Rajkovič [6] under the name DECMAK, already had all the main ingredients: tree-structured qualitative attributes, decision tables and decision rules, and algorithms supporting knowledge acquisition and explanation, including a graphical representation of decision tables and a machine-learning algorithm for constructing aggregate rules. About 30 practical applications, mainly in Slovenia, were reported at that time [6].

The name DEX (Decision EXpert) was first used in [7], to denote both the method and the supporting software that was developed at that time. In 2000, the DEX software was replaced by next-generation software called DEXi; at that point, the development team decided to keep the name DEX only for the method and use other names for its implementations.

DEX has always been closely tied with the supporting software. Due to the combinatorial nature of DEX's decision tables (explained in the next section), the method is unsuitable for manual construction of models and becomes practical only when supported by appropriate user interfaces and algorithms for knowledge elicitation, representation, verification, and explanation. In many aspects, the definition of the DEX method followed the actual software implementations, which is a somewhat unusual practice in the MCDM area.

Three generations of DEX-related software have been developed so far:

1. *DECMAK* [6] was released in 1981 for mini and personal computers under operating systems RT-11, VAX/VMS, and MS-DOS.
2. *DEX* [7] was released in 1987 as an integrated interactive computer program for VAX/VMS and MS-DOS.
3. *DEXi* [27] was released in 2000 for Microsoft Windows.

Originally, DEXi was designed as educational software (the letter "i" in DEXi, pronounced "ee", actually comes from the Slovenian "izobraževanje", education). DEXi was—and still is—used in Slovenian secondary schools and universities in MCDM and decision-support courses. Since 2000, additional features were gradually added to DEXi, which eventually became a complete, stable, and de facto standard implementation of DEX. DEXi supports an interactive creation and editing of all components of DEX models (attributes, their hierarchy and scales, decision tables, and alternatives) and provides methods for the evaluation and analysis of alternatives (what-if analysis, "plus-minus-1" analysis, selective explanation, comparison of alternatives). DEXi is free software, available at http://kt.ijs.si/MarkoBohanec/dexi.html together with other DEX-related software, which includes the following:

- *DEXiEval*, *JDEXi*, and *DEXi.NET*: Implementations of DEX evaluation procedure in different environments: command-line, Java, and C#, respectively,
- *DEXi HTML Evaluator*: A software package for running DEXi models in Web browsers, and
- *DEXx*: A Java-based software library [90].

3 Formal Representation of a DEX Model

The DEX method is defined from two aspects: static and dynamic. The static aspect gives a formal description of components and concepts of a DEX model. The dynamic aspect addresses algorithms and tools necessary to develop and modify the model and to use it for the evaluation and analysis of alternatives. In this section, we begin with static aspects and continue with dynamic aspects in the next. The formal notation is adapted from Trdin and Bohanec [90].

A DEX *model* M is a four-tuple $M = (X, D, S, F)$, where X is the set of *attributes*, S is the descendant function that determines the hierarchical structure of attributes, D is the set of value scales of attributes in X, and F is the set of aggregation functions.

3.1 Attributes

The set X contains n *attributes*: $X = \{x_1, x_2, \ldots, x_n\}$. Attributes are variables that represent observable properties of the decision problem and decision alternatives. In DEX models, attributes are usually given unique and meaningful names, such as *Price* and *Productivity*. In such cases, the notation x_i is conveniently and conventionally replaced by the corresponding attribute name.

3.2 Model Structure: Hierarchy of Attributes

Attributes in a DEX model are structured *hierarchically*. The structure is defined by the function $S : X \rightarrow 2^x$, which associates each $x \in X$ with a set of its *descendants* $S(x)$ in the hierarchy. The relation between an attribute and its descendants represents both dependence and influence: an attribute x *depends* on attributes in $S(x)$ and attributes from $S(x)$ *influence* x.

Given S, the set of *parents* of each $x \in X$ is defined as $P(x) = \{p \in X : x \in S(p)\}$. Attributes without parents are called *roots* and represent main *outputs* of the model. Attributes without descendants, $S(x) = \emptyset$, are called *basic* attributes and represent model *inputs*. Attributes with $S(x) \neq \emptyset$ are referred to as *aggregate* attributes and are also considered (partial, lower-level) *outputs* of the model.

The function S is required to represent a *hierarchy*, i.e., a connected and directed (from attributes to their descendants) acyclic graph with one or more roots. Figure 1 shows an example of a hierarchy, composed of ten attributes x_1 to x_{10} so that $S(x_1) = \{x_3, x_4\}$, $S(x_2) = \{x_5, x_6\}$, $S(x_4) = \{x_7, x_8, x_9\}$, $S(x_6) = \{x_9, x_{10}\}$, and $S(x_i) = \emptyset, i \in \{3, 5, 7, 8, 9, 10\}$. This means that x_1 and x_2 are roots. There are six basic attributes: x_3, x_5, x_7, x_8, x_9, and x_{10}. Among these, x_9 influences two parents, x_4 and

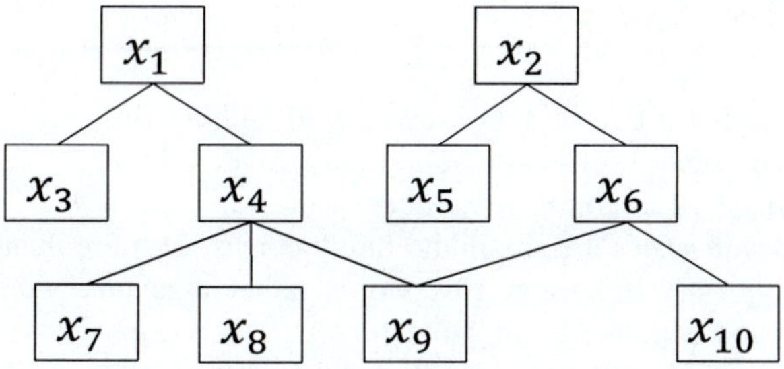

Fig. 1 Example of a hierarchy of attributes with two roots and 6 input attributes

x_6, while each of the remaining ones influences only one parent. Attributes x_1, x_2, x_4, and x_6 are aggregate and depend on their respective descendants.

In practice, DEX models are most often structured as trees rather than general hierarchies. A tree is a special type of hierarchy in which all attributes, except a single root attribute, have exactly one parent.

Example. Hereafter, we illustrate DEX concepts using a simple didactic model called *Employ*, which is distributed with the DEXi software. The model is aimed at the assessment of applicants for a Project Manager position in a small company. An earlier version was published in [8].

Figure 2 shows the structure of *Employ*. It consists of 12 tree-structured attributes. The root attribute is also called *Employ* and represents the output evaluation of job applicants. Applicants are assessed according to three groups of attributes, represented by aggregate attributes *Educat*, *Years,* and *Personal*. All of them are structured further, leading to seven basic attributes: *Formal*, *For.lang*, *Exper*, *Age*, *Comm*, *Leader*, and *Test* (see Fig. 2 for descriptions). These represent the observed characteristics of applicants and have the role of input variables.

Fig. 2 Structure of the *Employ* model with descriptions of attributes

Attribute	Description
Employ	Employee selection demo: Project manager
├**Educat**	Education
│ ├Formal	Formal education (degree)
│ └For.lang	Mastering of foreign language (English)
├**Years**	Age and experience
│ ├Exper	Professional experience in the field
│ └Age	Age of the candidate
└**Personal**	Personal characterisrics
├**Abilit**	Abilities
│ ├Comm	Cummunicability
│ └Leader	Leadership ability
└Test	Result of a psychological test

3.3 Scales

Each attribute $x \in X$ is associated with a value *scale* $D_x \in D$, which is defined as an ordered set of symbolic (qualitative) values: $D_x = \{v_{x,1}, v_{x,2}, \ldots, v_{x,m_x}\}$. Here, $m_x \geq 2$ denotes the number of discrete values that can be assigned to x. Usually, value scales are small and rarely consist of more than five values. Scale values are typically represented by words rather than numbers, for instance "low", "high", "unacceptable", and "good".

DEX scales can be either *ordered* or *unordered*.[1] Values of an ordered scale are assumed to be preferentially ordered so that $v_{x,1} \preccurlyeq v_{x,2} \preccurlyeq \cdots \preccurlyeq v_{x,m_x}$, where '$\preccurlyeq$' denotes a weak preference relation. Additionally, each scale D_x is partitioned in three subsets $B_x, N_x, G_x : B_x \cup N_x \cup G_x = D_x$, $B_x \cap N_x = B_x \cap G_x = N_x \cap G_x = \emptyset$. These subsets represent particularly bad, neutral, and particularly good values from D_x, respectively. They are convenient for displaying DEX values using different colors and fonts (usually red bold for **bad**and green bold italic for ***good*** values). By default, ordered scales are partitioned to $B_x = \{v_{x,1}\}$, $G_x = \{v_{x,m_x}\}$ and $N_x = D_x - (B_x \cup G_x)$, and unordered scales to $B_x = G_x = \emptyset$ and $N_x = D_x$.

According to the definition in [43], attributes that are associated with ordered scales are called *criteria*. In this way, a DEX model generally consists of attributes X, some of which are criteria. An attribute can be considered a criterion only after it has been associated with an ordered scale. For this reason, DEX is often referred to as a multi-attribute rather than multi-criteria method.

Example. Figure 3 shows the scales assigned to the attributes of *Employ*. The colors indicate that all scales, except D_{Age}, are ordered (increasingly) and partitioned using the default rule so that the worst and best attribute values appear at the beginning and end of the value list, respectively. Scale D_{Age} is unordered. Most of the values are represented by words: "unacc", "high", etc. Even though some values, for instance "1–5" and "21–25", are formulated as numeric intervals, they still represent single discrete symbols.

Fig. 3 Attributes of the *Employ* model associated with scales

Attribute	Scale
Employ	**unacc**; acc; good; ***exc***
├**Educat**	**unacc**; acc; ***good***
│ ├Formal	**prim-sec**; high; univ; MSc; ***PhD***
│ └For.lang	**no**; pas; ***act***
├**Years**	**unacc**; acc; ***good***
│ ├Exper	**no**; to1year; 1-5; 6-10; ***more***
│ └Age	18-20; 21-25; 26-40; 41-55; more
└**Personal**	**unacc**; acc; ***good***
├**Abilit**	**unacc**; acc; ***good***
│ ├Comm	**poor**; aver; good; ***exc***
│ └Leader	**less**; approp; ***more***
└Test	**D**; C; B; ***A***

[1] Actually, DEX implementations distinguish between *increasing*, *decreasing* and *unordered* scales. Here, we simplify the definition without loss of generality.

3.4 Aggregation Functions

The fourth and final component of the static DEX model definition is $F = \{f_x, x \in X\}$, a set of *aggregation functions* (also called *utility functions* in some software and older publications). An aggregation function serves for the evaluation of an aggregate attribute based on values of its immediate descendants in the model structure. Each aggregate attribute $y \in X$, $S(y) \neq \emptyset$ is thus associated with a total function.

$$f_y : D_{(1)} \times D_{(2)} \times \cdots \times D_{(k_y)} \rightarrow E_y,$$

where the Cartesian product refers to scales of $S(y) = \{x_{(1)}, x_{(2)}, \ldots, x_{(k_x)}\}$, where $x_{(1)}, x_{(2)}, \ldots, x_{(k_x)}$ are all descendants of y. In this way, $x_{(1)}, x_{(2)}, \ldots, x_{(k_x)}$ are arguments of f_y; in the context of f_y and corresponding decision tables, they are also referred to as *incoming* attributes. E_y denotes the range of f_y. Normally, the output range corresponds to the scale of y, that is, $E_y \equiv D_y$. However, for reasons that are explained later, E_y is often extended to:

- I_y, the set of intervals over D_y,
- $\mathcal{S}_y$, the power set of D_y,
- $\mathcal{P}_y$, probability distributions over D_y, or
- $\mathcal{F}_y$, fuzzy distributions over D_y.

In DEX, aggregation functions are represented by *decision tables*. Let us denote $C_y = D_{(1)} \times D_{(2)} \times \cdots \times D_{(k_x)}$ and $r_y = |C_y|$. Then, a decision table T_y consists of r_y entries $T_y = \{(\mathbf{x}_i, y_i), \mathbf{x}_i \in C_y, y_i \in E_y, i = 1, 2, \ldots, r_y\}$. Entries are often referred to as *elementary decision rules*: each rule defines the function value y_i for some combination of values of its arguments $\mathbf{x}_i$. Entries are required to be unique so that $\mathbf{x}_i \neq \mathbf{x}_j, i, j = 1, \ldots, r_y, i \neq j$. When completely defined, a decision table is normally expected to define output values for all possible $\mathbf{x} \in C_y$.

Example. Two completely defined decision tables are shown in Fig. 4. They define the functions that aggregate attributes *Abilit* and *Test* to *Personal* (left), and *Comm* and *Leader* to *Ability* (right). Each table contains 12 elementary decision rules, according

Fig. 4 Two decision tables, defining aggregation functions of *Personal* (left), and *Abilit* (right)

	Abilit	Test	Personal
1	**unacc**	**D**	**unacc**
2	**unacc**	C	**unacc**
3	**unacc**	B	**unacc**
4	**unacc**	***A***	**unacc**
5	acc	**D**	**unacc**
6	acc	C	**unacc**
7	acc	B	acc
8	acc	***A***	***good***
9	***good***	**D**	**unacc**
10	***good***	C	acc
11	***good***	B	***good***
12	***good***	***A***	***good***

	Comm	Leader	Abilit
1	**poor**	**less**	**unacc**
2	**poor**	approp	**unacc**
3	**poor**	***more***	**unacc**
4	aver	**less**	**unacc**
5	aver	approp	acc
6	aver	***more***	acc
7	good	**less**	**unacc**
8	good	approp	acc
9	good	***more***	***good***
10	***exc***	**less**	**unacc**
11	***exc***	approp	***good***
12	***exc***	***more***	***good***

to the number of possible value combinations of the corresponding C_y. Each value combination appears only once in each table. Each row in the table can be easily interpreted as an elementary "if–then" rule; for instance, rule 4 in the *Personal* table can be read as.

if *Abilit* = "unacc" and *Test* = "A" then *Personal* = "unacc".

In addition to two functions shown in Fig. 4, the *Employ* model contains three other decision tables, associated with attributes *Employ*, *Educat,* and *Years*; these are not shown here.

3.5 Alternatives

Once developed, a DEX model serves for the evaluation and analysis of decision alternatives. Formally, *alternatives* $\mathcal{A} = \{A_1, A_2, \ldots, A_q\}$ are not part of a DEX model M, but are rather considered as external data objects processed by M. Each *alternative* $A_i, i = 1, 2, \ldots, q$, is represented by a set of values:

$$A_i = \{a_{x,i} \in E_x, \forall x \in X\},$$

where each $a_{x,i}$ represents the value of A_i that is assigned to attribute x. Similarly as with aggregation functions, E_x is normally identical to D_x. However, in order to represent incomplete and/or uncertain information about alternatives, E_x may be in some contexts extended to value intervals, subsets, or value distributions.

The sets A_i are naturally partitioned in subsets I_i and O_i so that $A_i = I_i \cup O_i, I_i \cap O_i = \emptyset$. The two subsets correspond to basic and aggregate attributes of X, respectively. The former, I_i, represents basic observable properties of each A_i, which are defined by the decision maker and provide input data for the evaluation. In contrast, the values aligned with aggregate attributes, O_i, are calculated using the model and are thus obtained as results (outputs) of the evaluation. The most important results are those assigned to one or more roots of the model.

Example. In the *Employ* use case, alternatives are job applicants. Table 1 shows input data (that is, the corresponding B_i) of four applicants, named A, B, C, and D. In this case, all alternatives are represented by single values taken from the scales of corresponding attributes.

Table 1 Four job applicants, described by the values of basic attributes

Basic attribute	Applicants			
	A	B	C	D
Formal	MSc	*PhD*	*PhD*	*PhD*
For.lang	pas	*act*	*act*	*act*
Exper	to1year	*more*	6-10	6-10
Age	21-25	26-40	26-40	26-40
Comm	good	aver	good	*exc*
Leader	*more*	less	less	*more*
Test	B	B	C	*A*

3.6 Evaluation of Alternatives

Evaluation of alternatives is a process aimed at calculating output values of all alternatives that have been previously described by values of input attributes. Given some model M, the evaluation is carried out as a bottom-up aggregation of model inputs toward its outputs, according to the hierarchical structure of attributes and associated aggregation functions. Algorithmically, considering that a DEX model generally consists of a hierarchy of attributes, all attributes in M are first topologically sorted with respect to S. This determines the order of aggregation function evaluations and ensures the availability of all incoming inputs for calculating the output values of each subsequent aggregation function. Given function arguments, the output of that function is determined by a simple lookup in the corresponding decision table.

Example. Figure 5 shows evaluation results of the four applicants, defined previously in Table 1. Each column in Fig. 5 represents a complete set of values A_i of the corresponding applicant. The main outputs are assigned to the attribute *Employ*, indicating that the applicant D was assessed as "exc", A as "good", and the remaining two applicants as "unacc". Other outputs include values assigned to the remaining

Fig. 5 Evaluation of job applicants

Attribute	A	B	C	D
Employ	good	unacc	unacc	*exc*
├ **Educat**	acc	*good*	*good*	*good*
│ ├ Formal	MSc	*PhD*	*PhD*	*PhD*
│ └ For.lang	pas	*act*	*act*	*act*
├ **Years**	acc	*good*	*good*	*good*
│ ├ Exper	to1year	*more*	6-10	6-10
│ └ Age	21-25	26-40	26-40	26-40
└ **Personal**	*good*	unacc	unacc	*good*
├ **Abilit**	*good*	unacc	unacc	*good*
│ ├ Comm	good	aver	good	*exc*
│ └ Leader	*more*	less	less	*more*
└ Test	B	B	C	*A*

aggregate attributes *Educat*, *Years*, *Personal,* and *Abilit*. These values provide additional information about the candidates and help explaining the main results. For instance, one can easily see that the "unacc" *Personal* values of B and C have likely caused their "unacc" overall assessments, despite excellent assessments achieved at *Educat* and *Years*.

4 Dynamic Aspects of DEX

Dynamic aspects of DEX modeling refer to procedures, algorithms, and tools that are primarily used in two distinct decision analysis stages:

1. *Creation*: Here, the task is to develop an operational DEX model, usually starting from the scratch and aiming to satisfy both the goals of the decision maker and formal requirements, presented in the previous section. The main challenges addressed in this stage are how to (1) define the model and its components, (2) modify, edit, and maintain the model, (3) verify the model and its components (e.g., for completeness and consistency), (4) deal with uncertainty of knowledge and modeled phenomena, and (5) ensure transparency and comprehensibility of the model.
2. *Usage*: In this stage, one or more DEX models are already available and we want to use them to effectively solve the decision problem. This is associated with questions of how to (1) obtain and represent data about alternatives, (2) handle incomplete or uncertain data about alternatives, (3) evaluate alternatives, and (4) analyze, explain, justify, and validate results.

Among these, the representation and evaluation of alternatives have already been covered in the previous section. The remaining aspects are addressed in this section. The presentation is restricted to—and illustrated by—solutions implemented in the DEXi software.

4.1 Developing Model Components and Structure

DEX models are typically developed by individual *decision makers* or *groups*, the so-called *decision-problem owners* who are responsible for making the decision at hand. In the case of complex decision problems, the team is often extended with *experts* and *decision analysts*. The former provides expertise about the problem domain and help formulate model components. The latter are responsible for an appropriate use of the methodology and supporting tools and usually guide or even lead the process.

In most cases, DEX models are developed through *expert modeling*, i.e., "hand-crafting" of model components and structure, following the approach of expert systems. In this process, DEX models do not only "grow" from the scratch, but are often changed in other ways: attributes are added or deleted, their scales and

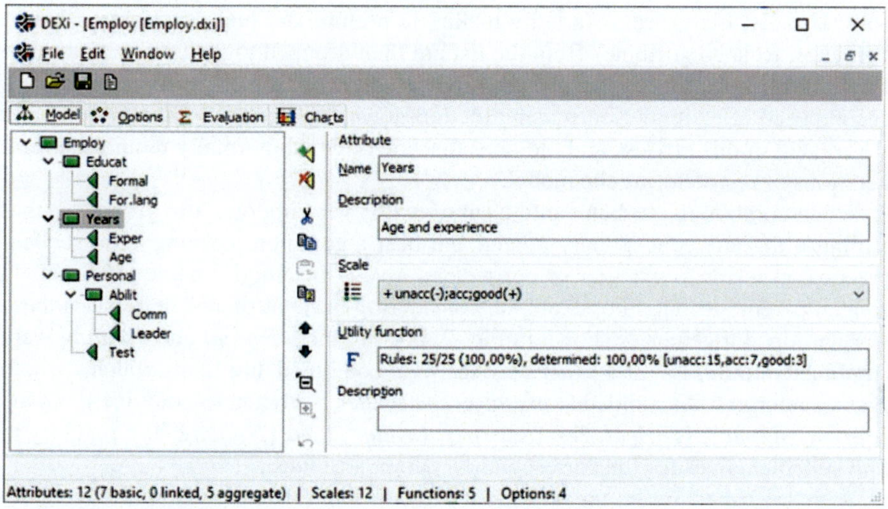

Fig. 6 DEXi model editor

aggregation functions are defined or changed, attribute hierarchies are restructured, etc. In practice, it is essential to support these needs by providing suitable software tools, such as DEXi.

For creating, editing and structuring attributes, DEXi provides an editor, shown in Fig. 6. All operations, mentioned above, are implemented, including model restructuring through drag-drop, duplicate, and copy-paste operations.

In addition to using software tools, many recommendations and "rules of thumb" of how to approach DEX modeling have been formulated from practical experience [27]. Regarding the selection of attributes, recommendations are the same as for any MCDM method: use attributes that are *relevant* for the problem and try not to overlook really *important* ones; avoid using *redundant* or closely correlated (*non-orthogonal*) attributes; assure that all input attributes are *operational* so that their values can be obtained for all alternatives in a sufficiently straightforward, well-defined, and accurate way.

With regard to developing model structure, DEX is similar to other hierarchical methods, such as AHP, but has some specific characteristics. In order to avoid too large decision tables (see the next section), it is recommended to make "narrow" hierarchies and limit the number of descendants of aggregate attributes to three or four at most. If an attribute requires, say, four descendants, consider structuring them further into sub-trees of $2 + 2$, $3 + 1$, or $2 + 1 + 1$ attributes.

In MCDM, two primary approaches are generally advocated for model structuring (see an overview in [58]): *top-down* (recursive decomposition of the root attribute to sub-trees) and *bottom-up* (defining input attributes first and gradually combining them toward the root). From practical experience, we can assure that none of them alone works really well; the most effective is the *middle-out* approach that combines both. Usually, the process starts by making a preliminary and unstructured list of attributes. Related attributes from the list are then grouped together in a bottom-up way, and attributes that seem too complex, too general, or too difficult to measure are decomposed into simpler ones using the top-down approach. Often, new attributes are created in this process and old ones discarded, which normally requires several iterations of restructuring the model.

When combining attributes into a subtree, it is very important to group together attributes that are conceptually related and bear a common meaning. An excellent practical criterion is whether or not we can give a meaningful name to the newly created parent attribute. For example, considering basic attributes in Table 1, these were grouped together as shown in Fig. 2. For instance, *Formal* and *For.lang* were combined into *Educat*, and *Exper* and *Age* were combined into *Years*, which are both easy to interpret. As a didactic exercise, the reader is invited to combine the pairs {*Formal*, *Exper*}, {*Formal*, *Test*}, {*Exper*, *Leader*}, and {*For.lang*, *Test*} and try to find suitable names for the corresponding parent attributes.

With regard to designing scales, the following recommendations have been formulated [27]:

- For basic attributes: use the least number of values that is still sufficient to distinguish between importantly different characteristics of alternatives with respect to that attribute. Usually, two to four values are sufficient. For instance, there are only three qualitatively different levels relevant to assess mastering of formal language (*For.lang*) in the *Employ* model: "no", "passive", and "active".
- For aggregate attributes: The number of values should gradually increase from input attributes toward the root. For example, three four-valued attributes might be aggregated into a five-valued attribute. Five-valued scales are generally recommended for root attributes, as they are usually sufficient and work quite well.
- On scale ordering: Use preferentially ordered scales whenever possible; they really help in the definition of decision tables. If some attribute does not have a natural preferential order, try reformulating or converting it to an ordered one. Avoid using decreasing scales; they tend to be less comprehensible than increasing ones.

4.2 Acquiring Decision Tables and Decision Rules

The evaluation process in DEX is guided by decision tables. In general, a decision table consists of elementary decision rules that determine output values for each combination of input values. This adds a combinatorial aspect and makes DEX decision tables somewhat harder to define than the corresponding aggregation functions in other MCDM methods, including AHP. In practice, it turned out that it was really important to provide interactive software tools that aid the development of decision tables.

Figure 7 illustrates three typical stages of creating a decision table in DEXi. The leftmost screenshot shows that DEXi automatically generates all possible value

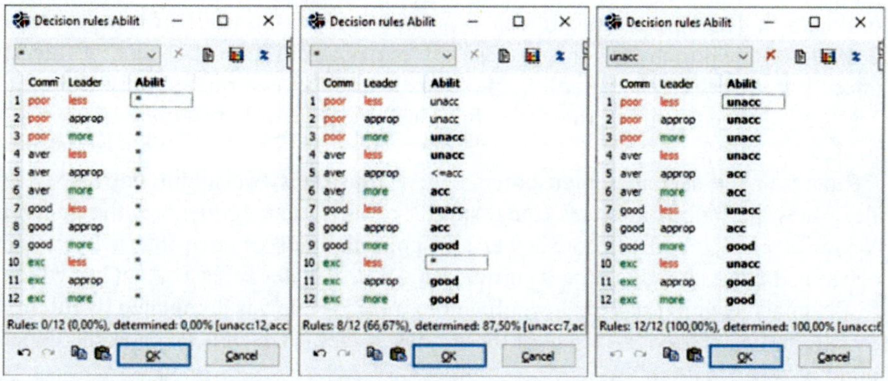

Fig. 7 Three stages of creating aggregation function for *Abilit* in DEXi

combination of descendant attributes (*Comm* and *Leader* in this example), releasing the decision maker from the burden of keeping track of all combinations. The right-most column initially contains asterisks '*', which indicate any possible value of the output attribute *Abilit*. It is important to understand that '*' represents the whole range of *Abilit*'s values, indicating that DEXi actually extends the notation E_y, introduced previously, to *intervals over* D_y. This extension is necessary for practical reasons and facilitates a smooth and user-friendly creation of decision tables from the scratch.

The second screenshot in Fig. 7 illustrates another important concept of DEX: considering the *principle of dominance* and trying to maintain the *consistency* of decision rules and *monotonicity* of aggregation function; for theoretical foundations, see [43, 44]. Let us assume that some decision table maps incoming attributes $x_1, x_2, \ldots, x_k$ to y, and all attributes are preferentially ordered. Suppose that a decision table already contains the entry

$$e = (\mathbf{x}_e, y_e), \mathbf{x}_e = \left(a_{1,e}, a_{2,e}, \ldots, a_{k,e}\right), a_{i,e} \in D_i, i = 1, 2, \ldots, k, y_e \in D_y.$$

Then, the principle of dominance requires that for any other entry f, where $\mathbf{x}_f \succcurlyeq \mathbf{x}_e$, it should hold $y_f \succcurlyeq y_e$ (and analogously for '$\preccurlyeq$'). Here, $\mathbf{x}_f \succcurlyeq \mathbf{x}_e$ is defined to hold if $a_{i,f} \succcurlyeq a_{i,e}$ for all $i = 1, 2, \ldots, k$, and $a_{i,f} \succ a_{i,e}$ is true for at least one i. In this case, f is said to *dominate* e. If none of the $\mathbf{x}_f \succcurlyeq \mathbf{x}_e$ or $\mathbf{x}_f \preccurlyeq \mathbf{x}_e$ are true, the entries e and f are incomparable. A decision table in which all comparable pairs of entries comply with the principle of dominance is *consistent* and defines a *monotone* aggregation function.

Even though one can define decision table entries one by one in succession, this is rarely done in DEXi because of the substantial help provided by the dominance principle. The second screenshot in Fig. 7 shows the situation where the decision maker has already defined eight entries: 3, 4, 6, 7, 8, 9, 11, and 12 (the respective output values are shown in bold). Comparing the entries 3 and 2, one can easily see that they differ only in the value of *Leader*. Since "more" $\succcurlyeq$ "approp", rule 3 dominates 2. The output value of rule 3 is "unacc", and the value of rule 2 should be worse or equal than that; this leaves only one possibility for the value of rule 2: "unacc". In this way, the value of rule 2 has been fully determined only from the previously defined value of rule 3. In this case, rule 3 provided an *upper bound* for the value of rule 2.

Rule 5 in the second screenshot in Fig. 7 illustrates two additional facts: (1) rule values are indeed intervals (the display "<= acc" actually denotes the interval ["unacc", "acc"]), and (2) both lower and upper bounds of such intervals can be determined from already defined entries. Rule 5 dominates rules 1, 2, and 4, which are all "unacc", which sets the lower bound of rule 5. Rule 5 is dominated by rules 6, 8, 9, 11, and 12. The worst value of these rules is "acc", which is taken as the upper bound of rule 5.

In this way, one can effectively develop a decision table by first providing a few entries, and then gradually assigning single values to entries that still contain intervals.

The third screenshot in Fig. 7 shows a fully developed table. If not overridden by the user, DEXi checks the consistency at all times and issues a warning if it is violated. Strictly following this procedure assures that the resulting tables (and consequently the whole model) are consistent and *complete*, i.e., they explicitly define output values for all possible combinations of input attribute values.

As already mentioned, DEX decision tables are sensitive to the number of incoming attributes and the size of their value scales: for k incoming attributes $x_1, x_2, \ldots, x_k$, the total number of entries equals to $r = \prod_{i=1}^{k} |D_i|$. In practice, it turns out that decision tables with sizes of up to 25 are small and usually quite easy to define. The difficulty increases toward the size of about 100, which is already quite difficult. Everything above 100 is very difficult, and everything above 500 is extremely hard if not impossible to define. The number of incoming attributes also matters: the more the attributes, the more difficult the rules to define, even if the size of the tables is comparable. In all such cases, it is strongly recommended [27] to restructure the model into narrower subtrees.

4.3 Restructuring Decision Tables

In some circumstances, it might be necessary to restructure the space around some already defined decision table, for instance by adding or deleting an incoming attribute or changing the definition of bounding scales. In practice, it is important to *preserve* as much information already contained in the table as possible. DEXi automatically restructures tables whenever possible. For example, Fig. 8 shows what happens with the table when the value "acc" is deleted from the scale of *Abilit*: the rules with previously assigned values "unacc" and "good" are preserved, and only previous "acc" entries need to be redefined.

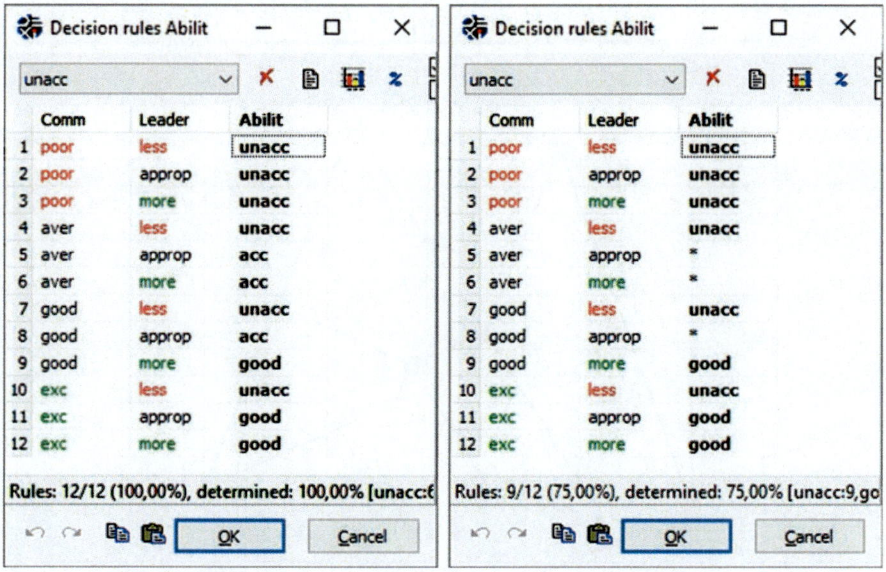

Fig. 8 Decision table *Abilit* before and after deleting "acc" from the output scale

4.4 *Representation of Decision Tables: Complex Rules and 3D Graphics*

Decision tables in DEXi are always acquired in terms of elementary decision rules (table rows). However, once completed, larger tables tend to become difficult to read and understand. To alleviate this problem, DEXi employs two methods: representation using complex rules and 3D graphic visualization.

The first method uses an algorithm that constructs a more compact table representation using *complex rules*. These are obtained by joining several elementary rules which have the same function value. The algorithm, whose presentation is beyond the scope of this chapter, belongs to the class of rule learning algorithms. Originally [6], it was adopted from the machine learning algorithm called AQ [59]. Recently, it has been enhanced for efficiency [51].

Using this algorithm, the *Abilit* decision table is presented in a more compact way with only 6 complex rules as shown in Fig. 9.

Fig. 9 Decision table *Abilit* represented with complex rules

	Comm	Leader	Abilit
1	poor	*	unacc
2	*	less	unacc
3	aver	>=approp	acc
4	aver:good	approp	acc
5	>=good	more	good
6	exc	>=approp	good

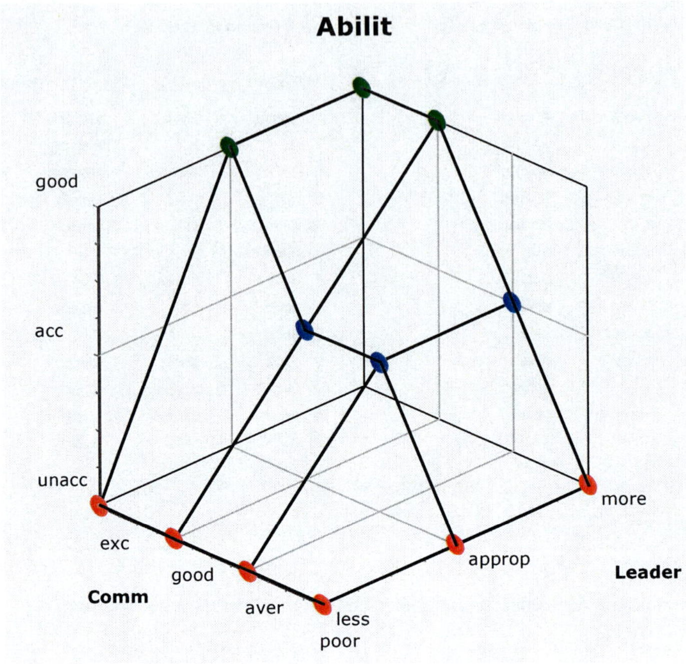

Fig. 10 Decision table *Abilit* represented with 3D graphic

The second method displays decision tables using 3D graphics (Fig. 10). There, table entries are interpreted as points in a multi-dimensional space. In the case of three or more incoming attributes, 3D intersections through the space are shown interactively. It is important to note that lines in Fig. 10 are there only to aid the 3D perception and are not part of the function definition, which remains discrete. It is also worth noticing that the function in Fig. 10 is somewhat typical for DEX; it resembles the minimum function and is not linear, in contrast with MCDM methods that use linear aggregation functions and weights.

4.5 Handling Incomplete Knowledge and Data

With this section, we turn attention to the usage stage, in which decision alternatives are represented and evaluated as described above in the formal section. In this stage, DEXi addresses two practically important issues: (1) handling incomplete data about alternatives and incompletely defined decision tables (this section) and (2) supporting analysis of the decision situation and individual alternatives (the next section).

As already indicated, DEX was inspired by ideas of expert systems. One of the most fundamental requirements for expert systems is that they must be able to process incomplete and uncertain knowledge. An expert system is expected to provide some answers, albeit incomplete or less accurate, even in the case of missing or uncertain input data, or "holes" in knowledge captured in the system.

The DEXi software implements a very simple version of this requirement using *value sets*: the notation E_x, introduced above, is extended to sets over D_x. In this way, values assigned to attributes by the evaluation algorithm are generally not single discrete values any more, but rather subsets of the corresponding scales. The evaluation algorithm iterates over all members of the input sets, and accumulates individual evaluations in corresponding output sets. Note that this approach handles both missing input data (which might be represented by '*', i.e., all values from the corresponding scale) and incompletely defined decision tables (by converting outgoing intervals to sets).

Figure 11 illustrates what happens in DEXi when some input data about job applicants is unknown. Candidate A has not been assessed with respect to his leadership abilities. Consequently, the model cannot really assess his *Personal* characteristics. The overall evaluation is represented by the set {"unacc", 'acc", "good"}, which does not say much, but indicates that A cannot reach the "exc" result. In contrast, candidates B and C are both assessed as "unacc", despite missing data of *Comm* and *For.lang*, respectively. Canididate D, whose *Test* results are currently unknown, achieved an extreme evaluation {"unacc", "exc"}. This indicates that she has the potential for becoming an excellent candidate, but subject to *Test* results, which may importantly determine the outcome.

Fig. 11 Evaluation of job applicants based on missing input data

Attribute	A	B	C	D
Employ	unacc; acc; good	unacc	unacc	unacc; ***exc***
├**Educat**	acc	***good***	*	***good***
│├Formal	MSc	***PhD***	***PhD***	***PhD***
│└For.lang	pas	***act***	*	***act***
├**Years**	acc	***good***	***good***	***good***
│├Exper	to1year	***more***	6-10	6-10
│└Age	21-25	26-40	26-40	26-40
└**Personal**	*	unacc	unacc	*
├**Abilit**	*	unacc	unacc	***good***
│├Comm	good	*	good	***exc***
│└Leader	*	less	less	***more***
└Test	B	B	C	*

4.6 Analysis of Alternatives

Analysis is one of the key concepts of MCDM and decision analysis in general. In contrast with evaluation, which merely calculates output results, analysis of alternatives is understood as an active involvement of participants who are trying to understand the decision situation, explain, and justify individual evaluations, explore the consequences of potential changes and search for better solutions. In DEXi, decision analysis is supported by three methods [27]: "what-if" analysis, selective explanation and "plus-minus-one" analysis.

What-if analysis is an exploration of consequences caused by changes of input data or aggregation functions. In DEXi, it is carried out through an iterative interactive process consisting of duplicating some alternative, changing data in one instance, and comparing both alternatives.

Selective explanation is aimed at the identification of particularly strong and weak characteristics of some alternative. Here, DEXi takes advantage of partitioning attribute scales into "good" and "bad" subsets. The method finds and displays all connected subtrees of attributes whose values are either all "good" or "bad". An example of such a display for job applicant B is shown in Fig. 12. It clearly highlights the candidate's main disadvantage, i.e., leadership abilities. On the other hand, the candidate does have advantages, reflected in *Educat* and *Exper*, so she might be considered for some other job position. Although based on a very simple idea, selective explanation has been found indispensable in practice for explaining and justifying decisions.

Plus-minus-one analysis investigates the effects of changing each basic attribute by one value down or up (if possible), independently of other attributes. Figure 13 shows results for candidate A. The column labeled A shows the current values, and the topmost value "good" is the current overall evaluation. The column "–1" shows the overall evaluation in the case that the corresponding attribute's value

Fig. 12 Job candidate B: Selective explanation

Weak points

Attribute	B
Employ	unacc
└**Personal**	unacc
├**Abilit**	unacc
└Leader	less

Strong points

Attribute	B
├**Educat**	***good***
├Formal	***PhD***
└For.lang	***act***
├**Years**	***good***
├Exper	***more***

Fig. 13 Job candidate A: Plus-minus-one analysis

Attribute	-1	A	+1
Employ		good	
├Formal		MSc	
└For.lang	unacc	pas	***exc***
├Exper	unacc	to1year	
└Age	unacc	21-25	
├Comm	acc	good	
└Leader	acc	***more***	]
└Test	acc	B	

drops by one. For instance, if *For.lang* were not "pas" but one step less (i.e., "no"), the candidate would have been evaluated as "unacc". In a similar way, the column " + 1" displays all possible improvements caused by one-step changes; it indicates that the candidate's evaluation may improve to "exc" if he improves his foreign language skills. Such displays require some practice to get used to, but effectively replace multiple "what-if" interactions.

5 Applications

The author of this chapter maintains a collection of DEX models that are available to him; they were developed mostly in the framework of various research and application projects, educational courses, or donated by other authors. In [22], he presented a study that included 582 models developed in 140 decision-making projects conducted in the period 1979–2015. Among these, 52 projects (38%) were documented in conference or journal publications, and further 20 (14%) projects were documented in internal reports. The collection is highly representative with respect to the addressed decision problems, decision makers involved, covered time period, and observed model characteristics.

The studied models addressed various decision problems from the following areas [22]:

- *Computer technology*: software, hardware, IT tools, programming languages, data base management systems, decision support systems;
- *Projects*: investments, research and R&D projects, tenders;
- *Organisations*: public enterprises, banks, business partners;
- *Schools*: quality of schools, programmes and teachers, school admission, choosing sports for schoolchildren;
- *Management*: production, portfolio management, trade, personnel (employees, jobs, teams), privatization, motorway;
- *Production*: location of facilities, technology, logistics, suppliers, office operations, construction, electric energy production, sustainability;
- *Ecology and Environment*: dumpsite/deposit assessment and remediation, emissions, ecological impacts, soil quality, ecosystem, sustainable development, protected areas;

- *Medicine and Health Care*: risk assessment (breast cancer, diabetes, ski injuries), nursing, technical analysis, knowledge management, healthcare network, therapy management for the Parkinson's disease and congestive heart failure;
- *Agriculture and Food Production*: economic and ecological effects of using genetically modified crops (GMOs), identification of (un)approved GMOs, coexistence of GMOs, crop protection, hop hybrids, garden quality;
- *Tourism*: nature trail, tourism farm facilities, mountain huts;
- *Services*: loans, housing loans, public portals, public services, leasing;
- *Other*: cars, hotels, electric motors, radars, game devices, awards, options, drug addiction, roof covering, coin design, data mining.

The study [22] also revealed some statistical properties of DEX models. An average model consists of roughly 28 attributes (16 of which are basic), 3.5 levels, and 2.5 descendants per aggregate attribute. The largest models may contain up to 400 attributes and 10 levels. An average scale contains 3.4 values and is preferentially ordered. An average decision table has 2.5 arguments, 3.7 output values, and 40 decision rules (with the median of 16). The overall completeness of decision tables is high (93%).

DEX applications generally belong to one of the following categories: (1) one-time decisions, (2) recurring decisions, and (3) decision support systems. These are reviewed next together with representative examples from the literature.

5.1 One-Time Decisions

Making *one-time decisions* is a classic MCDM task in which, given a set of decision alternatives, the goal is to choose the best alternative or to rank/sort them according to decision maker's preferences. Here, the main emphasis is on the quality of decision, i.e., trying to make the best possible decision in a given context. Consequently, the models tend to be very specific, they are often developed from the scratch or partly adapted from other sources, and they are quickly abandoned after the decision has been made.

First applications of DEX were mostly one-time and addressed decision problems related to computer technology, for instance choosing a data base management system [76] and purchasing a mainframe computer for a large factory [5]. The focus gradually shifted to other problem domains, such as employee selection [77]. Bohanec and Rajkovič [6] already report about 30 applications, including the selection of educational and production control software, microcomputers, as well as evaluation of trading partners, projects, and expert teams. Bohanec and Rajkovič [12] report on industrial applications, such as site suitability evaluation, product portfolio evaluation, and remediation of dumpsites. Similar problem types were approached ever since, for instance for evaluating public administration e-portals [57], project self-evaluation [99], mountain huts [88], and mountain lakes [79].

5.2 Recurring Decisions

Recurring decisions are essentially one-time decisions that occur periodically in similar circumstances, for instance, in approving loan applications or prescribing medical therapies. In this category, the emphasis shifts from the quality of individual decisions to the quality, generality, and usability of the model itself. The model has to "survive" multiple tries and be general enough to cope with changes from one case to another. The number of alternatives is initially unknown; sometimes, it may increase to hundreds or even thousands over time. This puts additional constraints on model design, which often proceeds by seeking the balance between including as many general attributes as possible (to facilitate considering cases that might emerge in the future) and reducing their number to only the most representative and easy to assess ones (to ease the burden of collecting input data for each considered alternative). Also, attributes and the whole decision-support procedure have to be clearly defined and meticulously documented, to prepare for multiple applications that may occur in long periods of time.

Since 1990s, with further development of supporting software, recurring decision problems became more and more accessible. Examples include supporting admission procedures in public schools [69], performance evaluation of enterprises [7], and evaluation of research and development projects [10]. Bohanec et al. [13] reported about recurring applications in health care in the assessment risks associated with breast cancer and diabetic foot. Probably the most important applications in the 1990s were *Talent*, a system for advising children in choosing sports [14], and a series of housing loan-allocation applications in collaboration with the Slovenian Housing Fund [11]. Both paved the way for decision support systems in the next period. More recent applications in recurring problems addressed, for instance, the evaluation of researchers [89], data mining workflows [100], detection of financial market manipulations [1], and water management investment projects [28].

5.3 Decision Support Systems

Many recurring decision problems look for the implementation of decision process in the form of a *decision support system* (DSS). DSSs are defined as interactive computer-based systems intended to help decision makers use communications technologies, data, documents, knowledge, and/or models to identify and solve problems, complete decision process tasks, and make decisions [72]. DEX models, developed for solving recurring problems, can be embedded in such DSSs in order to assess and analyze the given decision situations. DEX models usually provide just a fraction of the actual DSS functionality, which often adds a problem-specific user interface and includes additional support for user management, data acquisition, representation, search, and visualization, as well as other statistical, decision-analytic, and/or simulation methods.

Since 2005, many DSSs using DEX models were developed, most notably:

- *SMAC Advisor*: an advisory system on maize co-existence [16],
- *ESQI*: assessment of the impact of cropping systems on soil quality [17],
- a motorway traffic management system [70],
- *RIM*: assessment of bank reputational risk [20],
- *OVJE*: a DSS for the assessment of electric energy production technologies in Slovenia [23];
- *SIGMO*: assessment of GM presence in a food or feed products [24];
- *HeartMan*: a personal DSS for congestive heart management [25];
- *PD_manager*: a platform for Parkinson's disease management [91] with a DSS for the management of medication change [26, 63],
- *Soil Navigator*: assessment and management of soil functions [37],
- *IPSIM Chayote*: prediction and management of damage caused by fruit flies on the chayote in Reunion Island [38].

5.4 *Other Recent Applications*

Since 2005, DEX has been gaining more and more international reputation. It has been particularly well received in agronomy, agriculture, and related fields. Following a successful attempt of assessing economic and ecological impact of genetically modified crops [18, 98], a number of applications addressed the assessment of various cropping systems and their characteristics [4, 29, 33, 35, 36, 47, 54, 66, 71, 78, 80], production and marketing systems [32, 48, 55, 75, 83, 84], genetically modified crops [81, 92], farm management [67] and agri-food chains [61, 62].

Other recently conducted international applications of DEX addressed hydropower plant investments [87], assessment of offshore installation risks [41], employee redeployment [46], and development of ethno villages [74]. Ohunakin and Saracoglu [68] conducted a comparative study of methods MCDM, AHP, CDPC, DEX, ELECTRE III, and IV on the use case of very large concentrated solar power plants.

6 Two Real-World Examples

Among the above applications, we chose two for a more detailed showcase of the DEX approach and capabilities.

6.1 Example 1: Clay Pit Location

The first example came from the industry and was chosen because it represents a typical MCDM setting: a one-time decision problem aimed at choosing the best alternative from a given set. The problem was difficult and might have had critical consequences on the company and its long-term survival. Furthermore, initial alternatives were all unacceptable and better options had to be sought for in the process. The project was carried out in the 1990s; it is fully documented in the internal report [9] and partly in [12].

The company is called *Goriške opekarne* and is located near the Slovenian city of Nova Gorica. The company produces bricks and tiles. In 1993, they were faced with a difficult situation: the clay pit that had been providing raw material for their production became exhausted. The company had to find a replacement location, but this was difficult for a number of technological, logistic, financial, and environmental reasons, including a possible rejection of proposed solutions by local inhabitants. A group consisting of company managers, experts, and decision analysts was formed to define a DEX model and propose alternatives, while communicating with employees and inhabitants in a series of socio-psychological studies.

Eventually, a DEX model, whose complete structure is shown in Fig. 14, has been developed. A detailed description of individual attributes is beyond the scope of this chapter; however, one should note that the whole model is split in two main subtrees that address environmental and feasibility aspects of clay-pit locations, respectively. The model contains 30 basic and 19 aggregate attributes. Also, let us add that all scales in the model are preferentially ordered and the majority of them are either two-valued {"less-suit", "suit"} or three-valued {"unsuit", "less-suit", "suit"}. Scales of *ENVIRONMENT* and *ATTRACT* have four values, and the root attribute *SITE* has the scale {"unacc", "marg-acc", "less-acc", "acc", "good"}.

Decision rules from this model are illustrated here with just two examples shown in Fig. 15. The first example presents complex rules associated with attribute *TECH*, which aggregates three basic attributes: *TRANSPORT*, *CONSTRUCT,* and *LAND_ARCH*. TECH is located at the bottom of the tree; such attributes are often associated with specific decision rules and tables, which aim to resolve the decision problem at that level and provide useful evaluations/interpretations for higher levels of the model. The second example in Fig. 15 is located at the very root of the model and aggregates *ENVIRONMENT* and *FEASIBILITY* to the overall location evaluation (*SITE*). This is a typical representative of high-level aggregation functions, which tend to be symmetric or near-symmetric, and rule out all the cases that are evaluated poorly (i.e., "unacc") on lower levels of the hierarchy.

Three clay-pit locations were considered by this model: Okroglica, Marjetnica, and Bukovnik. Initially, all of them were assessed as "unacc". The team carried out a series of "what-if" scenarios, exploring possible improvements of the locations' characteristics, and anticipating an "optimistic" or "pessimistic" development of the investment project. Ultimately, eight variations were considered, which were

Attribute	Description
SITE	Site suitability
├**ENVIRONMENT**	Environmental components
│├**ATTRACT**	Site attractiveness
││├DEVELOP	Development factor
││├CHARACT	Site characteristics
││└**LAND**	Land
││ ├ACCESS	Land accessibility
││ ├QUALITY	Land quality (infrastructure)
││ └TIME_AVAIL	Time availability (short/long term)
│└**VULNER**	Site vulnerability
│ ├**POLLUTION**	Polution: Environmental impact
│ │├**LIV_ENV**	Pollution impacts to living environment
│ ││├**HUMAN**	Pollution impacts to humans
│ │││├HEALTH	Health impacts
│ │││└OTHER	Other impacts (e.g., noise)
│ ││├FAUNA	Pollution impacts on fauna
│ ││└FLORA	Pollution impacts on flora
│ │├SOC_ENV	Pollution impacts on social environment
│ │└**NON-LIV_ENV**	Pollution impacts on non-living environment
│ │ ├SOIL	Impacts on soil
│ │ ├WATER	Impacts on water
│ │ └AIR	Impacts on air
│ └**SITE_ORG**	Site organization
│ ├**VALUATION**	Site valuation
│ │├**ECOLOG**	Ecological valuation
│ ││├UNIQUE	Site uniqueness
│ ││└DIVERS	Site diversity
│ │└PERCEP	Perceptional valuation
│ └**USE**	Land use
│ ├DEMOGR	Demography
│ ├INFRAST	Infrastructure
│ └**POTENTIAL**	Site potentials
│ ├**PRIM_USE**	Primary use
│ │├AGRICULT	Agriculture
│ │├FOREST	Forestry
│ │└WATER	Water management
│ └**OTH_POT**	Other potentials
│ ├NAT_HER	Natural heritage
│ ├CULT_HER	Cultural heritage
│ └REC_TOUR	Recreation and tourism
└**FEASIBILITY**	Feasibility of the project
├SOC-PSYCH	Socio-psychological feasibility
├**TECH**	Technical feasibility
│├TRANSPORT	Transportation
│├CONSTRUCT	Construction
│└LAND_ARCH	Landscape architecture design
└**ECON**	Economic feasibility
├DIRECT	Direct expenses
└INDIRECT	Indirect expenses

Fig. 14 Structure of the *Clay Pit* DEX model

	TRANSPORT	CONSTRUCT	LAND_ARCH	TECH
1	*	*	**unsuit**	**unsuit**
2	**less-suit**	*	>=less-suit	less-suit
3	*	**less-suit**	>=less-suit	less-suit
4	*	*	less-suit	less-suit
5	***suit***	***suit***	***suit***	***suit***

	ENVIRONMENT	FEASIBILITY	SITE
1	**unacc**	*	**unacc**
2	*	**unacc**	**unacc**
3	less-acc	less-acc	marg-acc
4	less-acc	***acc***	less-acc
5	>=acc	less-acc	less-acc
6	acc	***acc***	acc
7	***good***	***acc***	***good***

Fig. 15 Two decision tables represented by complex rules: *TECH* and *SITE*

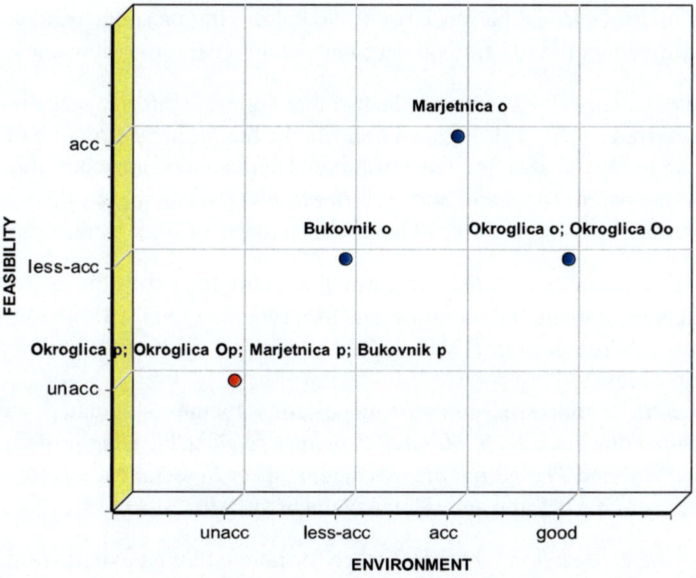

Fig. 16 Evaluation of Clay Pit locations along FEASIBILITY and ENVIRONMENT

evaluated as shown in the scatterplot in Fig. 16. Among these, "Marjetnica o" was considered the best and proposed for implementation.

6.2 *Example 2: Electric Energy Production Technologies*

The second example is taken from a more recent project aimed at the identification of reliable, rational, and environmentally sound production of electric energy in Slovenia by 2050 [23, 56]. Technology alternatives included both conventional and renewable energy sources: coal, gas, biomass, oil, nuclear, hydro, wind, and photovoltaic. This use case belongs to the category of complex and (potentially) recurring strategic decision problems, which occur and are relevant for any country. Without the ambition to go into any substantial detail, we wish to illustrate the capabilities of DEX to address really difficult real-world decision problems and handle models consisting of several tens of attributes, which are eventually incorporated in a DSS (called OVJE in this case, see above).

The methodological approach consisted of three stages, in which two DEX and one simulation model were developed:

- *DEX Model T* for the evaluation of eight electric energy production technologies.
- *DEX Model M* for the evaluation of mixtures of technologies, considering the shares of individual technologies in the total installed capacity.
- *Simulation Model S* for the evaluation of possible implementations of technology mixtures in the period 2014–2050, taking into account various scenarios of shutting down the existing power plants and constructing new ones.

Here, we shall briefly sketch only the first one; for more information, the interested reader is referred to [23, 56]. Figure 17 shows the hierarchical structure of Model T. There are 35 input and 28 aggregate attributes. There are two attributes that influence more than one parent (*Licences* and *Contribution to development*); therefore, this is a true hierarchy rather than a tree. The model consists of three main subtrees:

- *Rationality*: assesses how much a particular technology contributes to the overall societal development, the economy, and the prudent use of land with low pollution.
- *Feasibility*: addresses the *Technical*, *Economic,* and *Spatial* feasibility aspects of the technology.
- *Uncertainties*: addresses common uncertainty themes associated with energy policy and comprises *Technological dependence, Possible changes in society and in the world*, and *Perception of risks* with respect to technical advancement of a technology and trust into safety management system.

Among the 28 decision tables that were formulated by an expert team, we show here only two in the form of complex rules. Both tables are complete, consistent, and monotone. The first one (Fig. 18) aggregates the assessments of *Rationality, Feasibility,* and *Uncertainties* into the root assessment of the suitability of *Technology*. This table is *evaluative* because it evaluates some criterion (in this case *Technology*) according to evaluations of the incoming criteria: the better the value of each incoming criterion, the better the overall evaluation. Evaluative aggregation functions are typical for most MCDM methods.

The second table (Fig. 19) combines possible societal and world changes into a common perception of *Possible changes*. Here, the values "neg", "no", and "pos" refer to the direction of changes. Despite that one can assign preferences to these categories, they are not really evaluative. The table actually specifies a multi-variate *logic* for combining some basic concepts into higher-level concepts. This shows that in DEX, using multi-valued qualitative variables, it is possible to express both evaluative and logical rules. The latter usually occur at lower model levels and define concepts that enter the evaluation process at higher levels of the hierarchy. Inference based on logic is rarely featured in MCDM methods.

Using Model T, the study [23] concluded that there were only three technologies of sufficient suitability for Slovenia: Hydro, Gas, and Nuclear. Among these, Hydro is

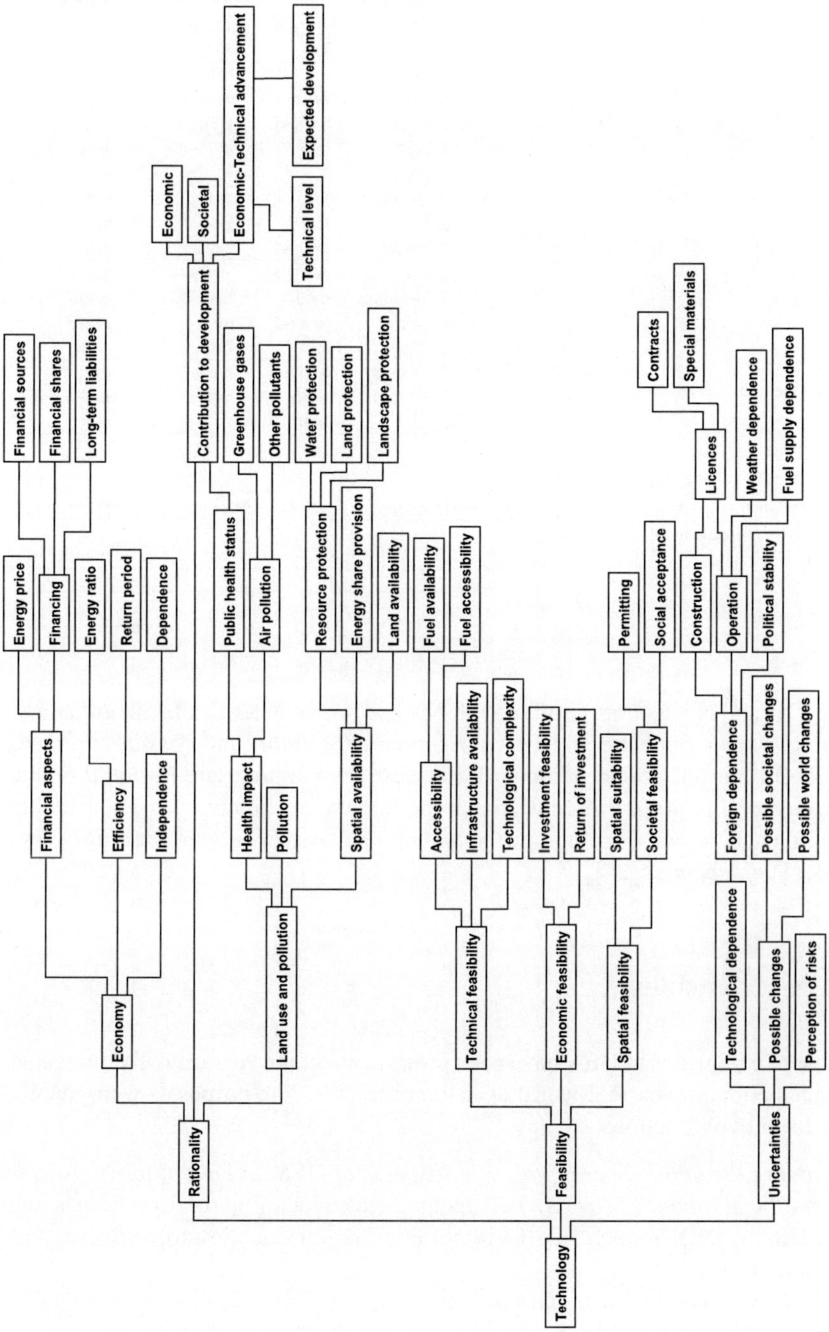

Fig. 17 Hierarchical structure of Model T

Fig. 18 Decision rules for the assessment of *Technology*

	Rationality	Feasibility	Uncertainties	Technology
1	inapprop	*	*	unsuit
2	<=low	<=med	v_high	unsuit
3	<=med	low	v_high	unsuit
4	>=low	low	high:med	weak
5	>=low	high	v_high	weak
6	>=med	>=med	v_high	weak
7	high	low	<=med	weak
8	high	*	v_high	weak
9	low:med	low	>=low	suit
10	>=low	low	low	suit
11	>=low	>=med	high	suit
12	low	>=med	>=med	good
13	low:med	med	med:low	good
14	>=low	>=med	med	good
15	high	low	none	good
16	>=med	>=med	none	exc
17	>=med	high	>=low	exc
18	high	>=med	>=low	exc

Fig. 19 Decision rules for determining the direction of *Possible* changes

	Possible societal changes	Possible world changes	Possible changes
1	neg	*	neg
2	*	neg	neg
3	no	no	no
4	>=no	pos	pos
5	pos	>=no	pos

the best. Gas and Nuclear are similar, with Nuclear worse in terms of *Feasibility* and *Perception of risks*, but better in terms of *Economic feasibility* and *Possible changes*. Coal and Oil are unsuitable particularly because of inappropriate *Rationality* due to *Land use and pollution*. All the remaining "green" technologies are unsuitable for a number of reasons, including *Economy*, *Land use*, *Economic feasibility*, and *Technological dependence*.

7 DEX Extensions

A number of extensions to DEX have been proposed over the years, mostly motivated by the needs of complex real-world decision problems. The proposals were mainly coming from two directions:

1. *Bridging the gap between qualitative aspects of DEX and quantitative aspects of the "traditional" MCDM*. This includes introducing numeric variables and weights in DEX models and facilitating numeric evaluation to better support ranking tasks.
2. *Taking advantage of artificial intelligence approaches*. This includes extended uncertainty handling mechanisms and using machine learning algorithms to

develop DEX models (semi)automatically from examples of past decisions, whenever such data is available.

7.1 Numeric Attributes

In its basic form, DEX is strictly qualitative. Currently, for instance, this requires that all numeric input data is pre-processed and discretized externally; introducing numerical variables to DEX models would definitely alleviate such problems and advance the generality of the approach, making it suitable for a larger class of problems. In principle, adding numerical attributes per se to the static formal model is easy, one should only extend the types of scales D. However, this is not enough because any such change should also preserve the dynamic aspects of the method: supporting the creation and modification of aggregation functions, considering completeness, consistency, and monotonicity of aggregation functions, and performing in the case of missing or uncertain data or knowledge. This is much harder and explains why the progress is slow and hesitant. Trdin and Bohanec [90] proposed a number of methodological extensions of this type, which will guide future evolution of DEX.

7.2 Weights

Traditional MCDM methods heavily rely on weights to define the importance of attributes [45]. The formal DEX model does not define any weights to be associated with qualitative attributes and decision rules. However, to bridge the gap between MCDM and also for practical reasons, DEX actually was extended with the notion of weights. The principle is simple:

- given a decision table that defines the function $y = f(x_1, x_2, \ldots, x_k)$ and consists of entries $(\mathbf{x}_e, y_e)$, $e = 1, 2, \ldots, r$,
- interpret the entries as points in a multi-dimensional space, and
- construct g as an approximation of f in the form

$$g(x_1, x_2, \ldots, x_k) = w_0 + w_1 \text{ord}(x_1) + \cdots + w_k \text{ord}(x_k).$$

Here, $\text{ord}(x)$ denotes the ordinal number of value x, and $w_i \in \mathcal{R}$ are relative weights of the corresponding arguments for $i = 1, \ldots, k$. These coefficients are determined using the least squares measure.

This method is actually implemented in DEXi and is used for approximate bi-directional transformations between weights and decision tables: (1) estimating weights from defined rules using the above approximation and (2) determining the values of yet undefined decision rules on the basis of already defined rules and user-specified weights. For more information, the reader is referred to [15, 27] and supplementary material in [38].

7.3 Combining Qualitative and Quantitative Evaluation

As already indicated, the qualitative foundation of DEX makes it particularly suitable for sorting and classification problems. In practice, however, it is sometimes necessary to use an already developed model also for ranking. For instance, whenever there are several alternatives assigned to the same evaluation category, it is often still necessary to tell them apart in some way. In qualitative DEX, this is in principle possible by refining the model by adding new categories and/or modifying decision rules to improve the separation; however, this requires redefining at least some parts of the model. Or alternatively, one can proceed by comparing similar alternatives, using analytic techniques to understand their advantages and disadvantages, and ranking them on this basis. In any case, both approaches are time consuming, and a better out-of-the-box support for ranking might alleviate such issues.

In principle, it is not difficult to think of some kind of numerical evaluation based on a DEX model. For instance, why not just taking the weights from the previous section and use the function g to carry out the calculations? Unfortunately, this does not work well because f and g might give different rankings based on the same inputs. The real challenge is how to assure that both evaluation procedures are *consistent* with each other. We are actually looking for a method that would first assign alternatives to distinct classes and only then rank them within each class. If possible, the process should not involve any additional work and should rely only on information already available in the model.

So far, there were two attempts at this kind of approach [8, 60]. They both explored the idea of representing values of some ordered attribute $x \in X$ in the form $v + \omega$, where $v \in D_x$ is a qualitative value of x, and $\omega \in [-0.5, +0.5]$ is a numerical *offset* to that value. The offset -0.5 is interpreted as "particularly bad" in the context of v, and $+0.5$ is interpreted as "particularly good". For instance, a job candidate evaluated as *Employ* = "good" + 0.33 would have been considered better than another candidate with *Employ* = "good"–0.12. In the evaluation algorithm, the qualitative evaluation of v remains exactly the same as before, and ω is assessed from the corresponding decision table using the principle of dominance and some additional assumptions. The approach of [8] uses a locally linear approximation of rules that map to some output category, whereas [60] uses copulas for the same purpose. The first approach is now called QQ (Qualitative-Quantitative). Unfortunately, these methods are not implemented in any currently available public software. We also think that the problem has not been solved in an entirely satisfactory way and remains a challenge for the future.

7.4 Handling Uncertainty Using Value Distributions

The idea of using fuzzy and probabilistic value distributions to cope with uncertain data and evaluations in DEX is actually quite old and originates from expert systems; it was first proposed in [5]. The idea is to allow using value distributions instead of single qualitative values in all places denoted E_x and E_y in the formal model. For instance, instead of assigning a single value to some input attribute, say *For.lang* = "pas", one can express their uncertainty about the real input using the probability distribution:

$$For.lang = \begin{pmatrix} "no" & "pas" & "act" \\ 0.1 & 0.7 & 0.2 \end{pmatrix}.$$

The same representation type can also be used for the outgoing values of decision rules.

This extension puts additional requirements on the evaluation procedure: the uncertainties, represented by probabilities or fuzzy possibilities, have to be propagated from input to output attributes in the hierarchy. Probabilistic inference employs product/sum operators, and fuzzy inference employs min/max or more general t-norm/t-conorm operators. For a more formal treatment of the subject, please see [90].

This evaluation procedure was actually implemented in the previous generation of DEX software and is still supported by software libraries *JDEXi, DEXi.NET,* and *DEXx*. It has been left out from DEXi for simplicity, but is destined to return in future software implementations.

7.5 Machine Learning of DEX Models

A large number of DEX application indicated that it is feasible for an individual decision maker or a group to develop a DEX model manually even for very difficult decision problems. On the other hand, it is also true that the task is demanding, particularly because the definition of decision rules generally requires more effort than definition of comparable aggregation functions in other MCDM methods. A natural question arising from DEX's artificial intelligence foundations is: could DEX models be constructed from data following the principles of machine learning? The answer is "yes, but it is hard"; none of the approaches attempted so far resulted in an entirely satisfactory solution for practice and no current general-purpose software implements any of the related methods.

The first and most ambitious attempt so far was made by Zupan et al. [95]. They proposed a method called HINT (Hierarchical Induction Tool) that is capable of transforming a large flat decision table into a hierarchical model, creating aggregate attribute and corresponding smaller decision tables along the way. This puts HINT in the category of concept learning methods [86]. Theoretically, the method did solve the task, but it also turned out very sensitive to noisy data (which is almost inevitable in practice) and required a very good coverage of the decision space by input data (which is also difficult to assure in practice).

The second attempt by Žnidaršič et al. [96, 97] was somewhat more modest and explored the approach of *model revision*: given an already developed DEX model and some data, the task is to revise model's decision rules so as to better match the data. Eventually, the method worked satisfactory, but its implementation *proDEX* [96] has become obsolete and is currently unsupported.

In the third attempt, [21] took an intermediate approach: given the *structure* of attributes and data, construct all aggregation tables in the model, taking into account probability distributions of input attributes and enforcing the principle of dominance. The authors demonstrated the approach by developing a model for predicting injury risk in ski resorts. The approach seems promising and will be further investigated in the future.

8 Summary

DEX is a qualitative decision modeling method that combines hierarchical and rule-based MCDM with artificial intelligence, specifically expert modeling and machine learning. The basic concepts of DEX are very simple and only involve hierarchically structured attributes, discrete scales, and decision tables consisting of elementary decision rules.

Despite simplicity, DEX has been successfully used in hundreds of real-world applications. According to its qualitative design, it is best suited for supporting sorting and classification decision problems. Choosing and ranking problems can be addressed, too, but they generally require some additional effort (interactive exploration and analysis of alternatives) or methodological extensions (such as QQ). Although DEX is suitable for one-time decision problems, recent trends indicate a shift toward recurring decision problems and including DEX models in DSSs. This is probably related with the effort that is required to develop a DEX model, which is generally greater than with comparable MCDM methods. One-time decision problems rarely justify the effort, whereas recurring and DSS ones do.

Practical applicability of DEX depends on the availability of supporting software. This is particularly true for the acquisition of decision tables, which might be very difficult on paper but becomes feasible when supported by appropriate tools and user interfaces. In addition to merely representing a static formal DEX model, DEX software always attempted to actively support dynamic aspects of creating and using the model. For DEX, it is really important to:

- facilitate *editing* of the model and its components: attributes, their structure, scales, aggregation functions, and alternatives;
- support the *acquisition* of decision rules, which includes enforcing the principle of dominance and checking for consistency and completeness at all times;

- maintain the *transparency* of the model and provide comprehensible representations of its components, such as complex rules and 3D graphics;
- provide various methods for the *analysis* of alternatives and *explanation* of evaluations.

DEX models may suffer from the combinatorial explosion: the size of decision tables increases exponentially with the number of incoming attributes. When developing a DEX model, it is thus important to follow recommendations that aim to keep the size below about 100: make "narrow" hierarchies with only 2 or 3 descendants of an aggregate attribute, and use the least number of values per attribute that still distinguishes between qualitatively different states of that attribute. Another potential disadvantage is that DEX, in its original form, is alien to numbers. When alternatives are prevalently described by numeric properties, the options are either to discretize them externally or use another MCDM method.

In the future, the main evolution will go in the direction of *Extended DEX*, as proposed by [90]. The proposal includes introduction of numeric attributes in DEX models and explicitly addressing uncertainty using probabilistic and fuzzy distributions of values. Software that partly supports these extensions already exists (DEXx software library), and full support is under development. The plan is to gradually replace the existing software DEXi with a new generation of web-based [52] and desktop applications. There also two challenges still open for further research and eventual software implementation: combined qualitative-quantitative evaluation of alternatives and learning DEX models from data.

Acknowledgements The author acknowledges the financial support from the Slovenian Research Agency, research core funding P2-0103.

References

1. Alić, I., Siering, M., Bohanec, M.: Hot stock or not? A qualitative multi-attribute model to detect financial market manipulation. In: Wigland, D.L. (Ed.) eInnovation: Challenges and Impacts for Individuals, Organizations and Society, Proceedings of 26th Bled eConference Bled, Slovenia, Kranj: Moderna organizacija, 64–77 (2013)
2. Bana e Costa, C., De Corte, J.-M., Vansnick, J.-C.: MACBETH (Overview of MACBETH multicriteria decision analysis approach). Int. J. Inf. Technol. Decis. Mak. **11**, 359–387 (2003)
3. Belton, V., Stewart, T.J.: Multi Criteria Decision Analysis: An Integrated Approach. Springer, US (2002)
4. Bergez, J.-E.: Using a genetic algorithm to define worst-best and best-worst options of a DEXi-type model: Application to the MASC model of cropping-system sustainability. Comput. Electron. Agric. **90**, 93–98 (2013)
5. Bohanec, M., Bratko, I., Rajkovič, V. (1983): An expert system for decision making. In: Sol, H.G. (Ed.) Processes and Tools for Decision Support, pp. 235–248, North-Holland.
6. Bohanec, M., Rajkovič, V. (1988): Knowledge acquisition and explanation for multi-attribute decision making. In: Proceedings of the 8th International Workshop Expert Systems and Their Applications AVIGNON 88, vol. 1, pp. 59–78, Avignon, 1988 (also available from http://kt.ijs.si/MarkoBohanec/pub/Avignon88.pdf)

7. Bohanec, M., Rajkovič, V.: DEX: an expert system shell for decision support. Sistemica **1**(1), 145–157 (1990)
8. Bohanec, M., Urh, B., Rajkovič, V.: Evaluating options by combined qualitative and quantitative methods. Acta Physiol. (Oxf) **80**, 67–89 (1992)
9. Bohanec, M., Kontić, B., Kos, D., Marušič, J., Polič, S., Rakovec, J., Sedej, B., et al. (1993): Comparison of clay-pit locations Okroglica, Bukovnik, Marjetnica with respect to environmental protection (in Slovene language). Ljubljana: Jožef Stefan Institute, Report DP-6742.
10. Bohanec, M., Rajkovič, V., Semolič, B., Pogačnik, A.: Knowledge-based portfolio analysis for project evaluation. Inf. Manag. **28**, 293–302 (1995)
11. Bohanec, M., Cestnik, B., Rajkovič, V.: Evaluation models for housing loan allocation in the context of floats. In: Berkeley, D., Widmeyer, G.R., Brezillon, P., Rajkovič, V. (eds.) Context Sensitive Decision Support Systems, pp. 174–189. Chapman & Hall, London (1998)
12. Bohanec, M., Rajkovič, V.: Multi-attribute decision modeling: industrial applications of DEX. Informatica **23**, 487–491 (1999)
13. Bohanec, M., Zupan, B., Rajkovič, V.: Applications of qualitative multi-attribute decision models in health care. Int. J. Med. Inf. **58–59**, 191–205 (2000)
14. Bohanec, M., Rajkovič, V., Leskošek, B., Kapus, V.: Expert knowledge management for sports talent identification and advising process. In: Carlsson, S.A., Brezillon, P., Humphreys, P., Lundberg, B.G., McCosh, A.M., Rajkovič, V. (eds.) Decision Support through Knowledge Management, pp. 46–59 (2000)
15. Bohanec, M., Zupan, B.: (2004): A function-decomposition method for development of hierarchical multi-attribute decision models. Decis. Support Syst. **36**, 215–233 (2004)
16. Bohanec, M., Messéan, A., Angevin, F., Žnidaršič, M.: SMAC advisor: a decision-support tool on coexistence of genetically-modified and conventional maize. In: Proceedings of Information Society IS 2006, Ljubljana, 9–12 (2006)
17. Bohanec, M., Cortet, J., Griffiths, B., Žnidaršič, M., Debeljak, M., Caul, S., Thompson, J., Krogh, P.H.: A qualitative multi-attribute model for assessing the impact of cropping systems on soil quality. Pedobiologia **51**, 239–250 (2007)
18. Bohanec, M., Messéan, A., Scatasta, S., Angevin, F., Griffiths, B., Krogh, P.H.: A qualitative multi-attribute model for economic and ecological assessment of genetically modified crops. Ecol. Model. **215**, 247–261 (2008)
19. Bohanec, M., Rajkovič, V., Bratko, I., Zupan, B., Žnidaršič, M.: DEX methodology: three decades of qualitative multi-attribute modelling. Informatica **37**, 49–54 (2013)
20. Bohanec, M., Aprile, G., Costante, M., Foti, M., Trdin, N.: A hierarchical multi-attribute model for bank reputational risk assessment. In: Phillips-Wren, G., Carlsson, S., Respício, A., Brézillon, P. (eds.) DSS 2.0 -- Supporting Decision Making with New Technologies. Amsterdam: IOS Press. ISBN 978–1–61499–398–8, 92–103 (2014).
21. Bohanec, M., Delibašić, B.: Data-mining and expert models for predicting injury risk in ski resorts. In: Decision Support Systems V - Big Data Analytics for Decision Making. First International Conference ICDSST 2015, pp. 46–60. Springer (2015)
22. Bohanec, M.: Multi-criteria DEX models (2017): an overview and analysis. In: Zadnik Stirn, L., et al. (eds.) SOR-2017: 14th International Symposium on Operational Research in Slovenia, Bled, Slovenia, September 27–29, 2017, pp. 155–160. Slovenian Society Informatika, Section for Operational Research, Ljubljana
23. Bohanec, M., Trdin, N., Kontić, B.: A qualitative multi-criteria modelling approach to the assessment of electric energy production technologies in Slovenia. CEJOR **25**, 611–625 (2017)
24. Bohanec, M., Mileva Boshkoska, B., Prins, T.W., Kok, E.J.: SIGMO: a decision support system for identification of genetically modified food or feed products. Food Control **71**, 168–177 (2017)
25. Bohanec, M., Tartarisco, G., Marino, F., Pioggia, G., Puddu, P.E., Schiariti, M.S., Baert, A., Pardaens, S., Clays, E., Vodopija, A., Luštrek, M.: HeartMan DSS: A decision support system for self-management of congestive heart failure. Expert Syst. Appl. **186**, 115688 (2021). https://doi.org/10.1016/j.eswa.2021.115688

26. Bohanec, M., Miljković, D., Valmarska, A., Mileva Boshkoska, B., Gasparoli, E., Gentile, G., Koutsikos, K., Marcante, A., Antonini, A., Gatsios, D., Rigas, F., Fotiadis, D.I., Tsiouris, K.M., Konitsiotis, S.: A decision support system for Parkinson disease management: Expert models for suggesting medication change. J. Decis. Syst. **27**, 164–172 (2018)
27. Bohanec, M.: DEXi: program for multi-attribute decision making, user's manual, version 5.04. IJS Report DP-13100, Jožef Stefan Institute, Ljubljana (available from http://kt.ijs.si/MarkoBohanec/pub/DEXiManual504.pdf) (2020)
28. Brelih, M., Rajkovič, U., Ružič, T., Rodič, B., Kozelj, D.: Modelling decision knowledge for the evaluation of water management investment projects. CEJOR **27**, 759–781 (2019)
29. Carpani, M., Bergez, J.-E., Monod, H.: Sensitivity analysis of a hierarchical qualitative model for sustainability assessment of cropping systems. Environ. Model. Softw. **27–28**, 15–22 (2012)
30. Chandra, V.: Artifical intelligence and machine learning. PHI Learning (2014)
31. Chen, S., Liu, J., Wang, H., Xu, Y., Augusto, J.C.: A linguistic multi-criteria decision making approach based on logical reasoning. Inf. Sci. **258**, 266–276 (2014)
32. Chopin, P., Tiroilien, J., Blazy, J.-M.: Ex-ante sustainability assessment of cleaner banana production systems. J. Clean. Prod. **139**, 15–24 (2016)
33. Colomb, B., Carof, M., Aveline, A., Bergez, J.-E.: Stockless organic farming: strengths and weaknesses evidenced by a multicriteria sustainability assessment model. Agron. Sustain. Dev. **33**, 593–608 (2012)
34. Corrente, S., Greco, S., Słowiński, R.: Multiple criteria hierarchy process in robust ordinal regression. Decis. Support Syst. **53**(3), 660–674 (2012)
35. Craheix, D., Bergez, J.-E., Angevin, F., Bockstaller, C., Bohanec, M., Colomb, B., Doré, T., Fortino, G., Guichard, L., Pelzer, E., Méssean, A., Reau, R., Sadok, W.: Guidelines to design models assessing agricultural sustainability, based upon feedbacks from the DEXi decision support system. Agron. Sustain. Dev. **35**, 1431–1447 (2015)
36. Craheix, D., Angevin, F., Doré, T., de Tourdonnet, S.: Using a multicriteria assessment model to evaluate the sustainability of conservation agriculture at the cropping system level in France. Eur. J. Agron. **767**, 75–86 (2016)
37. Debeljak, M., Trajanov, A., Kuzmanovski, V., Schröder, J., Sandén, T., Spiegel, H., Wall, D.P., Van de Broek, M., Rutgers, M., Bampa, F., Creamer, R.E., Henriksen, C.B.: A field-scale decision support system for assessment and management of soil functions. Front. Environ. Sci. **7**(115), 1–14 (2019)
38. Deguine, J.-P., Robin, M.-H., Corrales, D.C., Vedy-Zecchini, M.-A., Doizy, A., Chiroleu, F., Quesnel, G., Païtard, I., Bohanec, M., Aubertot, J.-N.: Qualitative modeling of fruit fly injuries on chayote in Réunion : development and transfer to users. Crop Prot. **129** (2021). ISSN: 105367-1-105367–11
39. Doumpos, M., Zopounidis, C.: Multicriteria Decision Aid Classification Methods. Springer, US (2002)
40. Efstathiou, J., Rajkovič, V.: Multiattribute decisionmaking using a fuzzy heuristic approach. *IEEE Transactions on Systems, Man, and Cybernetics,* SMC-9, 326–333 (1979)
41. Erdogan, G., Refsdal, A., Nygård, B., Petter Rosland, O., Kvam Randeberg, B.: Risk-based decision support model for offshore installations. Bus. Syst. Res. **9**(2), 55–68 (2018)
42. García-Lapresta, J.L., del Pozo, R.G.: An ordinal multi-criteria decision-making procedure under imprecise linguistic assessments. Eur. J. Oper. Res. **279**(1), 159–167 (2019)
43. Greco, S., Matarazzo, B., Słowiński, R.: Rough sets methodology for sorting problems in presence of multiple attributes and criteria. Eur. J. Oper. Res. **138**(2), 247–259 (2002)
44. Greco, S., Matarazzo, B., Słowiński, R.: Dominance-based rough set approach to decision under uncertainty and time preference. Ann. Oper. Res. **176**(1), 41–75 (2010)
45. Greco, S., Ehrgott, M., Figueira, J.: Multi Criteria Decision Analysis: State of the art Surveys. Springer, New York (2016)
46. Hajnić, M., Mileva Boshkoska, B.: A decision support model for the operational management of employee redeployment in large governmental organisations. J. Decis. Syst. (2020). https://doi.org/10.1080/12460125.2020.1768681

47. Hawes, C., Young, M.W., Banks, G., Begg, G.S., Christie, A., Iannetta, P.P.M., Karley, A.J., Squire, G.R.: Whole-systems analysis of environmental and economic sustainability in arable cropping systems: a case study. Agronomy **9**, 438 (2019)
48. Iocola, I., Campanelli, G., Diacono, M., Leteo, F., Montemurro, F., Persiani, A., Canali, S.: Sustainability assessment of organic vegetable production using a qualitative multi-attribute model. *Sustainability* 10(10), 3820 (2018). https://doi.org/10.3390/su10103820
49. Ishizaka, A., Nemery, P.: Multi-criteria Decision Analysis: Methods and Software. Wiley, Chichester (2013)
50. Jackson, P.: Introduction to Expert Systems, 2nd edn. Addison-Wesley (1990)
51. Kikaj, A., Bohanec, M.: Complex decision rules in DEX methodology: jRule algorithm and performance analysis. In: Proceedings of the 21th International Conference Information Society IS 2018, Volume A, Ljubljana: Jožef Stefan Institute, 17–20 (2018)
52. Kikaj, A., Bohanec, M.: DEX2Web—A web-based software implementing the multiple-criteria decision-making method DEX. In: Jayawickrama, U., Delias, P., Escobar, M.T., Papathanasiou, J. (eds.) Decision Support Systems XI: Decision Support Systems, Analytics and Technologies in Response to Global Crisis Management, pp. 30–43 (2021). https://www.springerprofessional.de/en/dex2web-a-web-based-software-implementing-the-multiple-criteria-/19166702
53. Kotsiantis, S.B., Zaharakis, I.D., Pintelas, P.E.: Machine learning: a review of classification and combining techniques. Artif. Intell. Rev. **26**, 159–190 (2007)
54. Maksimović, A., Grgić, Z., Puška, A., Šakić Bobić, B., Ćejvanović, F.: Application of multi-criteria decision making for the selection of apple varieties for the Northwestern region of B&H. J. Central Eur. Agric. **19**(3), 740–759 (2018)
55. Montemurro, F., Persiani, A., Diacono, M.: Environmental sustainability assessment of horticultural systems: a multi-criteria evaluation approach applied in a case study in mediterranean conditions. Agronomy **8**, 98 (2018). https://doi.org/10.3390/agronomy8070098
56. Kontić, B., Bohanec, M., Kontić, D., Trdin, N., Matko, M.: Improving appraisal of sustainability of energy options - A view from Slovenia. Energy Policy **90**, 154–171 (2016)
57. Leben, A., Kunstelj, M., Bohanec, M., Vintar, M.: Evaluating public administration e-portals. Inf. Polity **11**(3/4), 207–225 (2006)
58. MacSweeney Brugha, C.: Structure of multi-criteria decision-making. J. Oper. Res. Soc. **55**(11), 1156–1168 (2004)
59. Michalski, R.S., Larson, J.: Incremental generation of VLL hypotheses: the underlying methodology and the description of program AQ11. Report ISG 83–5, University of Illinois at Urbana-Champaign (1983)
60. Mileva Boshkoska, B., Bohanec, M.: A method for ranking non-linear qualitative decision preferences using copulas. Int. J. Decis. Support Syst. Technol. **4**(2), 42–58 (2012)
61. Mileva Boshkoska, B., Liu, S., Chen, H.: Towards a knowledge management framework for crossing knowledge boundaries in agricultural value chain. J. Decis. Syst. **27**, 88–97 (2018)
62. Mileva Boshkoska, B., Liu, S., Zhao, G., Fernandez, A., Gamboa, S., del Pino, M., Zarate, P., Hernandez, J., Chen, H.: A decision support system for evaluation of the knowledge sharing crossing boundaries in agri-food value chains. Comput. Ind. **110**, 64–80 (2019)
63. Mileva Boshkoska, B., Miljković, D., Valmarska, A., Gatsios, D., Rigas, G., Konitsiotis, S., Tsiouris, K.M., Fotiadis, D., Bohanec, M.: Decision support for medication change of Parkinson's disease patients. In: Computer Methods and Programs in Biomedicine, vol. 196, November 2020, 105552 (2020)
64. Mitchell, T.M.: Machine Learning. McGraw-Hill (1997)
65. Moshkovich, H.M., Mechitov, A.I.: Verbal decision analysis: foundations and trends. Adv. Decis. Sci. **2013**, 1–9 (2013)
66. Mouron, P., Heijne, B., Naef, A., Strassemeyer, J., Hayer, F., Avilla, J., Alaphilippe, A., Höhn, H., Hernandez, J., Mack, G., Gaillard, G., Solé, J., Sauphanor, B., Patocchi, A., Samietz, J., Bravin, E., Lavigne, C., Bohanec, M., Golla, B., Scheer, C., Aubert, U., Bigler, F.: Sustainability assessment of crop protection systems: sustainos methodology and its application for apple orchards. Agric. Syst. **113**, 1–15 (2012)

67. Nikoloski, T., Udovč, A., Pavlovič, M., Rajkovič, U.: Multi-criteria assessment model for farm reorientation. J. Decis. Syst. **27**, 79–87 (2018)
68. Ohunakin, O.S., Saracoglu, B.O.: A comparative study of selected multi-criteria decision-making methodologies for location selection of very large concentrated solar power plants in Nigeria. Afr. J. Sci. Technol. Innov. Dev. **10**(5), 551–567 (2018)
69. Olave, M., Rajkovič, V., Bohanec, M.: An application for admission in public school systems. In: Snellen, I.Th.M., van de Donk, W.B.H.J., Baquiast, J.-P. (eds.) Expert systems in public administration, Elsevier, pp. 145–160 (1989)
70. Omerčević, D., Zupančič, M., Bohanec, M., Kastelic, T.: Intelligent response to highway traffic situations and road incidents. In: Proceedings of TRA 2008, 21–24 April 2008, Ljubljana, pp. 1–6 (2008)
71. Pelzer, E., Fortino, G., Bockstaller, C., Angevin, F., Lamine, C., Moonen, C., Vasileiadis, V., Guérin, D., Guichard, L., Reau, R., Messéan, A.: Assessing innovative cropping systems with DEXiPM, a qualitative multi-criteria assessment tool derived from DEXi. Ecol. Ind. **18**, 171–182 (2012)
72. Power, D.J.: Decision Support, Analytics, and Business Intelligence, 2nd edn. Business Expert Press, New York (2013)
73. Puppe, F.: Systematic Introduction to Expert Systems: Knowledge Representations and Problem-Solving Methods. Springer, Berlin (1993)
74. Prevolšek, B., Maksimović, A., Puška, A., Pažek, K., Žibert, M., Rozman, Č.: Sustainable development of ethno-villages in Bosnia and Herzegovina—a multi criteria assessment. Sustainability **12**, 1399 (2020). https://doi.org/10.3390/su12041399
75. Prišenk, J., Rozman, Č., Pažek, K., Turk, J., Bohak, Z., Borec, A.: A multi-criteria assessment of the production and marketing systems of local mountain food. Renew. Agric. Food Syst. 1–10 (2013)
76. Rajkovič, V., Bohanec, M. (1980): A cybernetic model of the computer aided decision making. In: Proceeding of 9th Internatioal Congress on Cybernetics, pp. 185–199, Namur
77. Rajkovič, V., Bohanec, M., Batagelj, V.: Knowledge engineering techniques for utility identification. Acta Physiol. (Oxf) **683**(1–3), 271–286 (1988)
78. Ravier, C., Prost, L., Jeuffroy, M.-H., Wezel, A., Paravano, L., Reau, R.: Multi-criteria and multi-stakeholder assessment of cropping systems for a result-oriented water quality preservation action programme. Land Use Policy **42**, 131–140 (2015)
79. Ravnikar, T., Bohanec, M., Muri, G.: Monitoring and assessment of anthropogenic activities in mountain lakes: A case of the Fifth Triglav Lake in the Julian Alps. Environ. Monit. Assess. **188**(4), 1–17 (2016)
80. Rezaei, M.E., Barmaki, M., Veisi, H.: Sustainability assessment of potato fields using the DEXi decision support system in Hamadan Province, Iran. J. Integr. Agric. **17**(11), 2583–2595 (2018)
81. Ricci, B., Messéan, A., Lelièvre, A., Colénod, F.-C., Angevin, F.: Improving the management of coexistence between GM and non-GM maize with a spatially explicit model of cross-pollination. Eur. J. Agron. **77**, 90–100 (2016)
82. Roy, B.: Paradigms and challenges. In: Greco, S., Ehrgott, M., Figueira, J. (eds.) Multi Criteria Decision Analysis: State of the Art Surveys. Springer, New York (2016)
83. Rozman, Č, Grgić, Z., Maksimović, A., Ćejvanović, F., Puška, A., Šakič Bobić, B.: Multiple-criteria approach of evaluation of milk farm models in Bosnia and Herzegovina. Mljekarstvo **66**(3), 206–214 (2016)
84. Rozman, Č, Maksimović, A., Puška, A., Grgić, Z., Pažek, K., Prevolšek, B., Ćejvanović, F.: The use of multi criteria models for decision support system in fruit production. Erwerbs-obstbau **59**, 235–243 (2017)
85. Saaty, T.L., Vargas, L.G.: Models, Methods, Concepts & Applications of the Analytic Hierarchy Process. Springer, US, New York (2012)
86. Sammut, C.: Concept learning. In: Sammut, C., Webb, G.I. (eds.) Encyclopedia of Machine Learning. Springer, Boston, MA (2010). https://doi.org/10.1007/978-0-387-30164-8_154

87. Saracoglu, B.: A qualitative multi-attribute model for the selection of the private hydropower plant investments in Turkey. J. Ind. Eng. Manag. **9**(1), 152–178 (2016)
88. Stubelj Ars, M., Bohanec, M.: Towards the ecotourism: a decision support model for the assessment of sustainability of mountain huts in the Alps. J. Environ. Manag. **91**(12), 2554–2564 (2010)
89. Taškova, K., Stojanova, D., Bohanec, M., Džeroski, S.: A qualitative decision-support model for evaluating researchers. Informatica **31**(4), 479–486 (2007)
90. Trdin, N., Bohanec, M.: Extending the multi-criteria decision making method DEX with numeric attributes, value distributions and relational models. CEJOR **26**, 1–41 (2018)
91. Tsiouris, K.M., Gatsios, D., Rigas, G., Miljković, D., Koroušić-Seljak, B., Bohanec, M., Arredondo, M.T., Antonini, A., Konitsiotis, S., Koutsouris, D.D., Fotiadis, D.I.: PD_manager: an mHealth platform for Parkinson's disease patient management. Healthc. Technol. Lett. **4**(3), 102–108 (2017)
92. Wohlfender-Bühler, D., Feusthuber, E., Wäger, R., Mann, S., Aubry, S.J.: Genetically modified crops in Switzerland: implications for agrosystem sustainability evidenced by multi-criteria model. Agron. Sustain. Dev. **36**, 33 (2016)
93. Zadeh, L.A., Klir, G.J., Yuan, B. (eds.) Fuzzy Sets, Fuzzy Logic, and Fuzzy Systems: Selected Papers by Lotfi A Zadeh. WSPC (1996)
94. Zimmermann, H.-J.: Fuzzy Sets, Decision Making, and Expert Systems, 4th edn.International Series in Management Science Operations Research, vol. 10. Springer, Netherlands (2001)
95. Zupan, B., Bohanec, M., Demšar, J., Bratko, I.: Learning by discovering concept hierarchies. Artif. Intell. **109**, 211–242 (1999)
96. Žnidaršič, M., Bohanec, M., Zupan, B.: proDEX–A DSS tool for environmental decision-making. Environ. Model. Softw. **21**(10), 1514–1516 (2006)
97. Žnidaršič, M., Bohanec, M.: Automatic revision of qualitative multi-attribute decision models. Found. Comput. Decis. Sci. **32**(4), 315–326 (2007)
98. Žnidaršič, M., Bohanec, M., Zupan, B.: Modelling impacts of cropping systems: demands and solutions for DEX methodology. Eur. J. Oper. Res. **189**, 594–608 (2008)
99. Žnidaršič, M., Bohanec, M., Lavrač, N., Cestnik, B.: Project self-evaluation methodology: the healthreats project case study. In: Proceedings of Information Society 2009, Ljubljana, pp. 85–88 (2009)
100. Žnidaršič, M., Bohanec, M., Trdin, N.: Qualitative assessment of data-mining workflows. In: Respício, A., Burstein, F. (eds.) Fusing Decision Support Systems into the Fabric of the Context, pp. 75–88. IOS Press, Amsterdam (2012)

Analysis of Fuzzy AHP and Fuzzy TOPSIS Methods for the Prioritization of the Software Requirements

Mohd. Nazim, Chaudhary Wali Mohammad, and Mohd. Sadiq

Abstract Software requirement prioritization is a key activity of elicitation process whose objective is to select the top requirements based on the ranking values for the implementation. Different methods have been proposed to prioritize the software requirements using various techniques like AHP, TOPSIS, etc. under fuzzy environment. The objective of this chapter is to compare two multi-criteria decision making methods, i.e., fuzzy AHP and fuzzy TOPSIS, for the prioritization of the software requirements. The experimental work is carried out on ten functional requirements and three non-functional requirements of an Institute Examination System.

Keywords Software requirement prioritization · Fuzzy AHP · Fuzzy TOPSIS · Agreement measure · Institute examination system

Abbreviations

AHP	Analytic Hierarchy Process
FDM	Fuzzy Decision Matrix
FR	Functional Requirement
FUSE	Fuzzy Synthetic Extent
IES	Institute Examination System
MCDM	Multi-Criteria Decision Making
NFR	Non-functional Requirement
NIS	Negative Ideal Solution

Mohd. Nazim · C. Wali Mohammad
Computer Science and Technology Research Group, Department of Applied Sciences and Humanities, Faculty of Engineering and Technology, Jamia Millia Islamia, (A Central University), New Delhi 110025, India
e-mail: cmohammad@jmi.ac.in

Mohd. Sadiq (✉)
Department of Computer Science and Automation, Indian Institute of Science, Bangalore 560012, India
e-mail: msadiq@jmi.ac.in

A. J. Kulkarni (ed.), *Multiple Criteria Decision Making*, Studies in Systems, Decision and Control 407, https://doi.org/10.1007/978-981-16-7414-3_4

PIS	Positive Ideal Solution
SR	Software Requirement
TOPSIS	Technique for Order of Preference by Similarity to Ideal Solution
$\sum$:	Summation
$\otimes$:	Fuzzy multiplication operator

1 Introduction

Stakeholders play an important role during the software requirements (SRs) elicitation process because they are the main sources of the SRs. A system is successful if it has been developed according to the need of the stakeholders. Different requirements elicitation techniques are employed to identify the SRs like "traditional methods", "group elicitation methods", etc. There are hundreds or thousands of requirements after the completion of the SRs elicitation process. These requirements are mainly divided into "functional requirements" (FRs) and "non-functional requirements" (NFRs) [1, 2]. In real life applications, all the elicited SRs cannot be implemented due to different constraint of an organization like time limitations, resource limitation, finance, etc. Therefore, it is an important issue that how to prioritize the SRs when different stakeholders participate during the decision making process.

The SRs prioritization is a multi-criteria decision making (MCDM) approach whose objective is to prioritize the SRs based on their ranking values [1, 3]. In literature, different methods have been developed to prioritize the SRs using crisp and fuzzy data like "Analytic Hierarchy Process" (AHP), "Technique for Order of Preference by Similarity to Ideal Solution" (TOPSIS) [4]. AHP is based on pairwise comparison of decisions among FRs and NFRs so that high ranked FRs can be identified [5]. In the case of TOPSIS method, the selection is performed on the basis of "positive ideal solution" (PIS) and the "negative ideal solution" (NIS) of the FRs [6]. Both methods work on the crisp data, but in real-life applications, the data may be inadequate and vague. Therefore, to deal this issue, both the methods have been applied successfully under fuzzy environment in different areas like management science, software engineering, etc. In literature less attention is given on the comparative study between fuzzy AHP and fuzzy TOPSIS based on the agreement measure matric. In this chapter an attempt has been made to present a comparison of the accuracy of the fuzzy AHP and fuzzy TOPSIS methods using SRs prioritization problem.

The rest part of this chapter is organized as follows: The related work in the area of SRs prioritization is discussed in Sect. 2. An insight into fuzzy AHP and fuzzy TOPSIS methods are given in Sect. 3. The experimental work is carried out in Sect. 4. Finally, the conclusion and suggestion for future work are given in Sect. 5.

2 Related Work

SRs prioritization and selection is an important research problem in the area of software engineering and information systems. Different methods have been proposed to compute the ranking values of the SRs during the SRs prioritization process like AHP, TOPSIS, Planning game, etc. These methods have been compared on different systems based on different criteria. For example, Karlsson et al. [7] compared six prioritization methods by using a project of sixteen quality requirements for a mobile phone system. Three decision makers were invited during the evaluation process. Based on the comparative study, the authors found that AHP is most favorable method for the prioritization of the SRs. A similar experiment was conducted in [8] by considering the five SRs prioritization methods, in which fourteen decision makers have participated for the prioritization of the thirteen requirements. In another study, Karlsson et al. [9] performed an experimental work based on two SRs prioritization methods, i.e., pair-wise comparisons and planning game partitioning. Perini et al. [10] compared the accuracy of "*AHP and CBRanking techniques*" in the area of SRs prioritization. Among various MCDM methods, fuzzy AHP and fuzzy TOPSIS methods have received much attention in the area of facility location selection, supplier selection, evaluation of the business intelligence vendors, etc. [11–15]. Ertugrul and Karakasoglu [11] compared "fuzzy AHP and fuzzy TOPSIS methods" for facility location selection problem. Alavi et al. [12] applied these two methods for plant species selection. Junior et al. [13] focused on the comparison between fuzzy AHP and TOPSIS in the area of supplier selection. As per our knowledge there is no study in literature which compares the fuzzy AHP and fuzzy TOPSIS methods for the SRs prioritization problem.

3 An Insight into Fuzzy AHP and Fuzzy TOPSIS Methods

This section presents a brief discussion on fuzzy AHP and fuzzy TOPSIS methods.

3.1 Fuzzy AHP

The AHP was proposed by Thomas L. Saaty as an MCDM tool to capture the expert's knowledge. In traditional AHP, exact numbers are used and it cannot be used to deal with the vagueness and imprecision during the decision making process. To overcome this problem, fuzzy AHP was developed to solve the hierarchical problems. The fuzzy AHP is an MCDM method which has been used for the selection and prioritization of the alternatives in different area of management science and engineering [5]. In this chapter, the extent fuzzy AHP is employed for SRs prioritization, which was introduced by Chang in 1996 [16].

Let $FR = \{fr_1, fr_2, \ldots, fr_p\}$ and $NFR = \{nfr_1, nfr2_2, \ldots, nfr_q\}$ be functional requirements and non-functional requirements, respectively. According to [16], each FR is used and extent analysis is carried out for each NFR, respectively. So, for each FR the q number of "extent analysis values" with the following signs is obtained as follows:$E_{bi}^1, E_{bi}^2, \ldots, E_{bi}^p; i = 1, 2, \ldots, p$, where$E_{bi}^j, j = 1, 2, 3, \ldots\ldots, q$ are TFNs. The value of the fuzzy synthetic extent (FUSE) with respect to the ith FR is defined as follows:

$$FUSE_i = \sum\nolimits_{j=1}^{q} E_{bi}^j \otimes \left[\sum_{i=1}^{p}\sum_{j=1}^{q} E_{bi}^j\right]^{-1} \tag{1}$$

To find out $\sum_{j=1}^{q} E_{bi}^j$, the process of fuzzy addition of q extent analysis values for a specific matrix is performed as follows:

$$\sum_{j=1}^{q} E_{bi}^j = \left(\sum_{j=1}^{q} l_j, \sum_{j=1}^{q} m_j, \sum_{j=1}^{q} u_j\right). \tag{2}$$

To find out $\left[\sum_{i=1}^{p}\sum_{j=1}^{q} E_{bi}^j\right]^{-1}$, the fuzzy addition operation of $E_{bi}^j (j = 1, 2, \ldots, q)$ values is performed as follows:

$$\sum_{i=1}^{p}\sum_{j=1}^{q} E_{bi}^j = \sum_{i=1}^{p} l_i, \sum_{i=1}^{p} m_i, \sum_{i=1}^{p} u_i. \tag{3}$$

The inverse of the above is computed as follows:

$$\left[\sum_{i=1}^{p}\sum_{j=1}^{q} E_{bi}^j\right]^{-1} = \left(\frac{1}{\sum_{i=1}^{p} u_i}, \frac{1}{\sum_{i=1}^{p} m_i}, \frac{1}{\sum_{i=1}^{p} l_i}\right). \tag{4}$$

Let $E_1 = (l_1, m_1, u_1)$ and $E_2 = (l_2, m_2, u_2)$ are two TFNs, then the degree of possibility of $E_2 \geq E_1$ is defined as follows:

$$P(E_1 \geq E_2) = \sup_{y \geq x} \lfloor \min(\mu_{E_1}(x), \mu_{E_2}(y)) \rfloor,$$

$$P(E_1 \geq E_2) = 1, \text{iff} m_1 \geq m_2,$$

$$P(E_2 \geq E_1) = hgt(E_1 \cap E_2)\mu_{E_1}(od)$$

Here, od is the ordinate of the maximum intersection point between the membership function of E_1 and E_2. The ordinate od is given by.

$$P(E_2 \geq E_1) = \frac{l_1 - u_2}{(m_2 - u_2) - (m_1 - l_1)} \tag{5}$$

The values of $P(E_1 \geq E_2)$ and $P(E_2 \geq E_1)$ will be used to compare two fuzzy numbers, i.e., E_1 and E_2. The degree of possibility for a convex fuzzy number to be greater than d convex fuzzy numbers E_i $(i = 1, 2, \ldots, d)$ can be defined as follows:

$$P(E \geq E_1, E_2, E_3, \ldots.., E_d) = P[(E \geq E_1) and (E \geq E_2) and \ldots and (E \geq E_d)].$$

Let $od(X_i) = \min P(FUSE_i \geq FUSE_d)$ for $d = 1, 2, \ldots, p; d \neq i$. Then the weight vector (V') can be defined as follows:

$$V' = \left((od'(X_1)), od'(X_2), \ldots.., od'(X_p)\right)^T$$

where $X_i (i = 1, 2, \ldots, p)$ are p FRs. After normalization, the normalized weight vectors are defined as follows:

$$V = \left((od(X_1)), od(X_2), \ldots.., od(X_p)\right)^T \tag{6}$$

where,

V is a non-fuzzy number.

3.2 *Fuzzy TOPSIS*

The TOPSIS method was proposed by Hwang and Yoon as an MCDM method for the selection and prioritization of the alternatives [17]. The main idea of this method is that the selected FRs and NFRs should have little distance from the positive ideal solution and the large distance from negative ideal solutions. The objective of positive ideal solution (PIS) is to minimize the cost of the SRs and maximizes the benefit, whereas the negative ideal solution (NIS) maximizes the cost and minimizes the benefit. Following steps of the fuzzy TOPSIS method have been used for the prioritization of the SRs: (a) identify decision makers, (b) find out the FRs and NFRs of a system, (c) find out the linguistic variables that will be used for the evaluation of the FRs and NFRs, (d) aggregate the weight of the NFRs, (e) construct the fuzzy decision matrix (FDM), (f) normalize the FDM, (g) construct normalized weighted FDM, (h) compute fuzzy PIS and fuzzy NIS, (i) compute the distance of each FR from fuzzy PIS and NIS, and (j) compute the closeness coefficients of each FR [18, 19]. Both extent fuzzy AHP [16] and fuzzy TOPSIS [18, 19] methods were implemented using Python programming language. The comparative study between these two methods based on the SRs is given in the next section. The notion of the fuzzy logic was developed by Lotfi A. Zadeh to deal with vagueness and imprecision during the decision making process [20].

4 Experimental Work

In this chapter we have considered a small dataset for the prioritization of the SRs of an Institute Examination System (IES) [21]. To compute the ranking values of the FRs of an IES, we have implemented both fuzzy AHP and fuzzy TOPSIS methods using Python programming language. The experiments were carried out on Python 3.8.1, the JetBrains PyCharm Professional 2019, Intel(R) Core(TM) i3-6006U CPU @ 2.00 GHz, 4.00 GB RAM, and 64-bit Operating System. We performed an experiment of prioritizing the requirements on a dataset of ten FRs and three NFR of an IES. The list of the FRs and NFRs of an IES includes the following: FR1:"the printout of the bank receipt of student's fee", FR2:"entry of the internal and external marks of the student", FR3:"view the result of the semester", FR4:"generate the seating arrangement for the examination", FR5:conduct the online examination", FR6:"examination form filling", FR7:"upload any other activity related to examination", FR8: "issue the admit card for the examination", FR9:"provide the approval for the examination form", and FR10:"online payment of the fee for the examination", NFR1: "security", NFR2: "cost, and NFR3: "usability" [21]. Five decision makers were invited to perform the evaluation of ten FRs.

We have designed two tests, i.e., Test-1 and Test-2, for the comparison between fuzzy AHP and fuzzy TOPSIS methods. The meaning of Test-1 and Test-2 is given below:

- Test-1: Both fuzzy AHP and fuzzy TOPSIS methods have used the same values of the NFRs
- Test-2: Both fuzzy AHP and fuzzy TOPSIS methods have used the distinct values of the NFRs

Based on our analysis, we observed that in case of fuzzy AHP, both the test produces the same ranking order of the FRs of an IES, i.e., $FR_1 > FR_{10} > FR_9 > FR_6 > FR_4 > FR_3 > FR_7 > FR_2 > FR_8 > FR_5$, as shown in Fig. 1. But there was some difference in the ranking order in case of fuzzy TOPSIS method, i.e., T1:$FR_1 > FR_6 > FR_4 > FR_9 > FR_{10} > FR_7 > FR_8 > FR_2 > FR_3 > FR_5$ and T2: $FR_6 > FR_1 > FR_{10} > FR_4 > FR_9 > FR_8 > FR_2 > FR_7 > FR_3 > FR_5$. The visual representations of fuzzy TOPSIS method based on Test-1 and Test-2 are exhibited in Fig. 2.

The ranking orders of the FRs in the outputs of the programs for both the tests (i.e., Test-1 and Test-2) by using the fuzzy AHP and the fuzzy TOPSIS methods are shown in Table 1 and Table 2, respectively. The agreement measure metric has been employed to compute the difference in the ranking orders of the FRs produced by fuzzy AHP and fuzzy TOPSIS methods [10]. The agreement measure for both the tests is exhibited in Fig. 3. In Test-1, it is observed that for the 1st, 5th, 9th, and 10th positions the agreement measure is maximum, i.e., 1.0, while it is minimum (i.e., 0.33) for the 3rd position. The agreement measures are very close to each other for the 6th, 7th, and 8th positions, i.e., 0.83, 0.85, and 0.87, respectively, as shown in Fig. 3a. Similarly, the result of Test-2 is exhibited in Fig. 3b. The comparison of

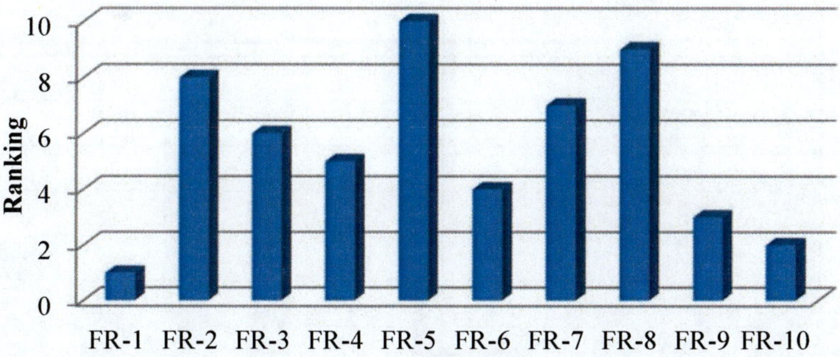

(a) Test-1

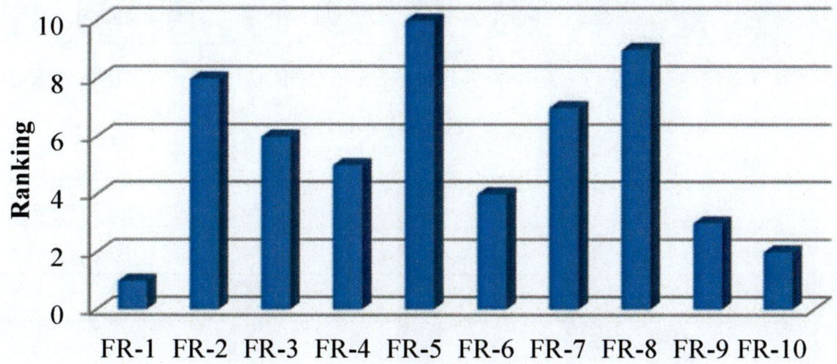

(b) Test-2

Fig. 1 The ranking of FRs by using fuzzy AHP method

fuzzy AHP and the fuzzy TOPSIS for the different parameters is illustrated in Table 3.

5 Conclusions and the Future Work

In this chapter we have presented the comparison between two MCDM methods under fuzzy environment, i.e., fuzzy AHP and fuzzy TOPSIS. Both fuzzy AHP and fuzzy TOPSIS methods were implemented using Python language for the computation of the ranking values of the FRs of an IES. The agreement measure was used to compare the ranking order between fuzzy AHP and fuzzy TOPSIS. In our experimental work, we have conducted two tests based on the inputs. As a result, it was observed that both tests produced the same results when fuzzy AHP was used for computing the ranking values of the FRs. In case of the fuzzy TOPSIS method, there were some differences in the ranking order. We have used the agreement measure metric to compute the differences in the ranking order of fuzzy AHP and fuzzy

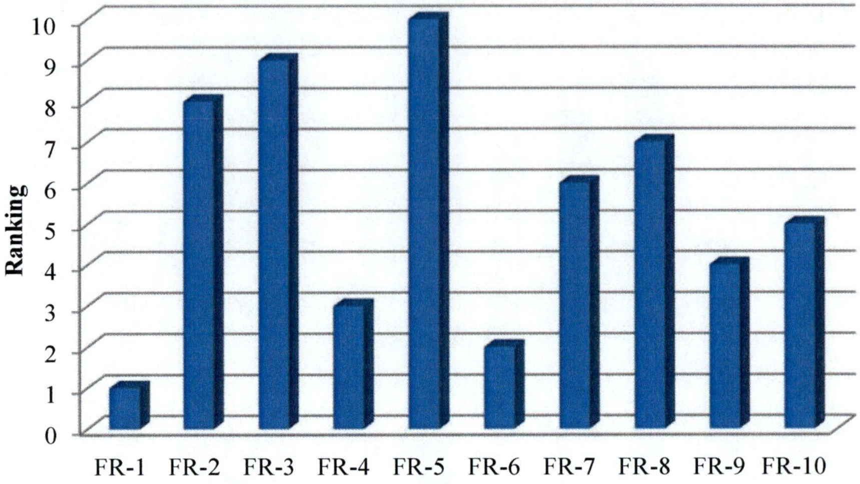

(a) Test-1

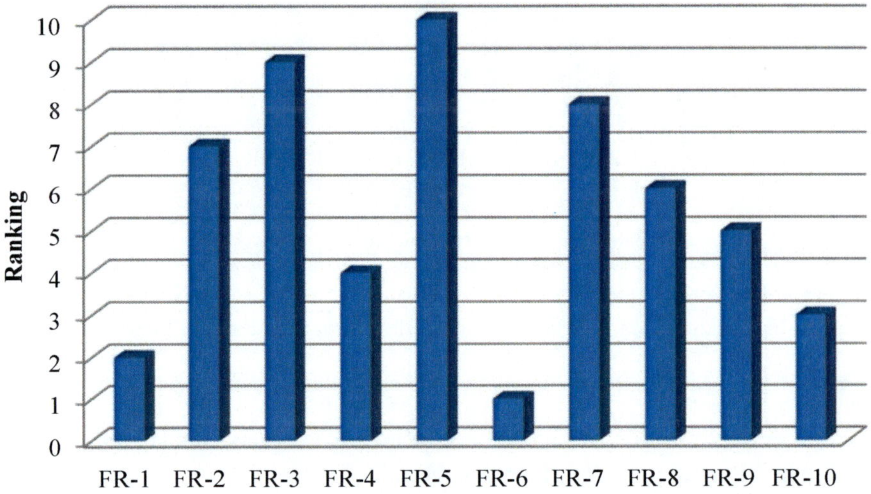

(b) Test-2

Fig. 2 The ranking of FRs by using fuzzy TOPSIS method

TOPSIS methods. One of the limitations of this study is that small set of the FRs and NFRs have been used in the experimental work; and for the analysis, only two methods have been used. In future, we will try to compare more than two methods and will analyze the SRs by considering the large dataset of SRs.

Table 1 Program's output for the ranking of FRs by using fuzzy AHP method

Ranking	Functional requirements	
	Test-1	Test-2
Rank-1	FR_1	FR_1
Rank-2	FR_{10}	FR_{10}
Rank-3	FR_9	FR_9
Rank-4	FR_6	FR_6
Rank-5	FR_4	FR_4
Rank-6	FR_3	FR_3
Rank-7	FR_7	FR_7
Rank-8	FR_2	FR_2
Rank-9	FR_8	FR_8
Rank-10	FR_5	FR_5

Table 2 Program's output for the ranking of FRs by using fuzzy TOPSIS method

Ranking	Functional requirements	
	Test-1	Test-2
Rank-1	FR_1	FR_6
Rank-2	FR_6	FR_1
Rank-3	FR_4	FR_{10}
Rank-4	FR_9	FR_4
Rank-5	FR_{10}	FR_9
Rank-6	FR_7	FR_8
Rank-7	FR_8	FR_2
Rank-8	FR_2	FR_7
Rank-9	FR_3	FR_3
Rank-10	FR_5	FR_5

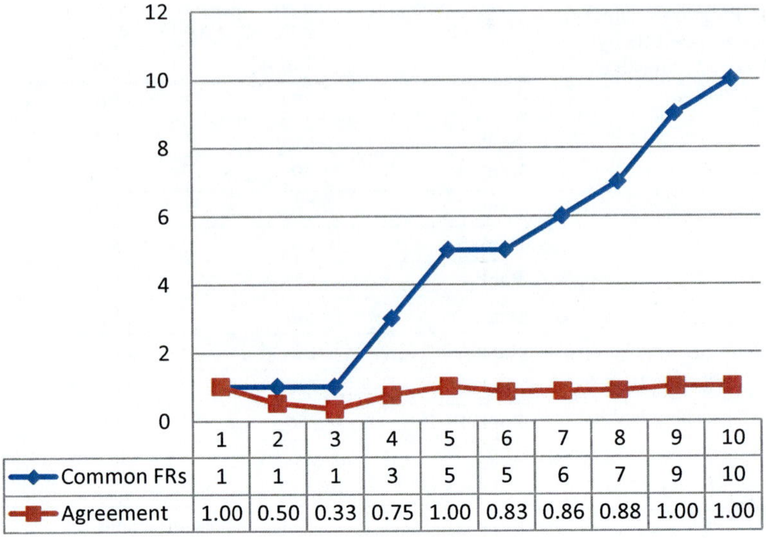

(a) Test-1

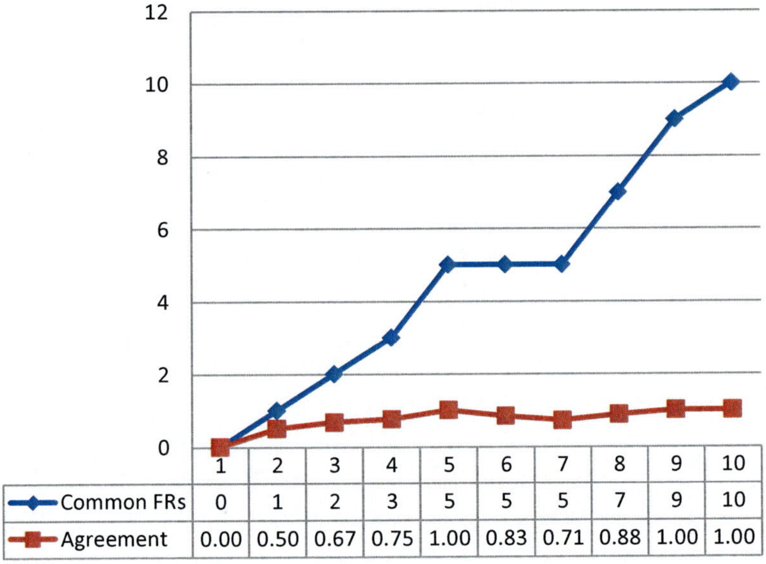

(b) Test-2

Fig. 3 Agreement measure

Table 3 Comparison between fuzzy AHP and fuzzy TOPSIS methods

S. No	Description	Test No	MCDM Methods	
			Fuzzy AHP	Fuzzy TOPSIS
1	Ranking of FRs	Test-1	Same	Different
		Test-2		
2	Best FR	Test-1	FR_1	FR_1
		Test-2	FR_1	FR_6
3	Worst FR	Test-1	FR_5	FR_5
		Test-2	FR_5	FR_5
4	No. of positions of FRs for which agreement measures are maximum	Test-1	4 (i.e., 1st, 5th, 9th, 10th)	
		Test-2	3 (i.e., 5th, 9th, 10th)	
5	No. of positions of FRs for which agreement measures are half	Test-1	1(i.e., 2nd)	
		Test-2	1(i.e., 2nd)	
6	No. of positions of FRs for which agreement measures are very close to each other	Test-1	3(i.e., 6th, 7th, 8th)	
		Test-2	2(i.e., 6th, 8th)	
7	No. of positions of FRs for which agreement measures are minimum	Test-1	1 (i.e., 3rd)	
		Test-2	1 (i.e., 1st)	

References

1. Sadiq, M., Jain, S.K.: Applying fuzzy preference relation for requirements prioritization in goal oriented requirements elicitation process. Int. J. Syst. Assur. Eng. Manag. **5**, 711–723 (2014)
2. Sadiq, M., Nazneen, S.: Elicitation of software testing requirements from the selected set of software's requirements in GOREP. Int. J. Comput. Syst. Eng. Indersci. **5**(3), 152–160 (2019)
3. Nazim, M., Mohammad, C.W., Sadiq, M.: Generating datasets for software requirements prioritization research. In: 2020 IEEE International Conference on Computing, Power and Communication Technologies (GUCON), Greater Noida, India, pp. 344–349 (2020)
4. Sadiq, M., Khan, S., Mohammad, C.W.: Selection of software requirements using TOPSIS under fuzzy environment. Int. J. Comput. Appl. 1–10 (2020)
5. Laarhoven, P.J.M.V., Pedrycz, W.: A fuzzy extension of Saaty's priority theory. Fuzzy Sets Syst. **11**(1–3), 229–241 (1983)
6. Chen, C.T.: Extensions of the TOPSIS for group decision-making under fuzzy environment. Fuzzy Sets Syst. **114**, 1–9 (2000)
7. Karlsson, J., Wohlin, C., Regnell, B.: An evaluation of methods for prioritizing software requirements. Inf. Softw. Technol. **39**(14–15), 939–947 (1998)
8. Ahl, V.: An experimental comparison of five prioritization methods: investigating ease of use, accuracy and scalability. Master's thesis, School of Engineering. Blekinge Institute of Technology. Sweden (2005)
9. Karlsson, L., Thelin, T., Regnell, B., Berander, P., Wohlin, C.: Pair-wise comparisons versus planning game partitioning - experiments on requirements prioritisation techniques. Empir. Softw. Eng. **12**(1), 3–33 (2007)
10. Perini, A., Susi, A., Ricca, F., Bazzanella, C.: An empirical study to compare the accuracy of AHP and CBRanking techniques for requirements prioritization. In: Fifth International Workshop on Comparative Evaluation in Requirements Engineering, pp. 23–35. New Delhi (2007)

11. Ertugrul, I., Karakasoglu, N.: Comparison of fuzzy AHP and fuzzy TOPSIS methods for facility location selection. Int. J. Adv. Manuf. Technol. **39**, 783–795 (2008)
12. Alavi, I., Alinejad-Rokny, H.: Comparison of fuzzy AHP and fuzzy TOPSIS methods for plant species selection (case study: reclamation plan of Sungun Copper Mine; Iran). Aust. J. Basic Appl. Sci. **5**(12), 1104–1113 (2015)
13. Junior, F.R.L., Osiro, L., Carpinetti, L.C.R.: A comparison between fuzzy AHP and fuzzy TOPSIS methods to supplier selection. Appl. Soft Comput. **21**, 194–209 (2014)
14. Soloukdar, A., Parpanchi, S.A.: Comparing fuzzy AHP and fuzzy TOPSIS for evaluation of business intelligence vendors. Decis. Sci. Lett. **4**(2), 137–164 (2015)
15. Ouma, Y.O., Opudo, J., Nyambenya, S.: Comparison of fuzzy AHP and fuzzy TOPSIS for road pavement maintenance prioritization: methodological exposition and case study. Adv. Civil Eng. 17 (2015)
16. Chang, D.Y.: Applications of the extent analysis method on fuzzy AHP. Eur. J. Oper. Res. **95**, 649–655 (1996)
17. Hwang, C.L., Yoon, K.: 1981. Multiple Attribute Decision Making. Lecture Notes in Economics and Mathematical Systems, vol. 186. Springer, Berlin (1981)
18. Chen, C.T.: Extensions of the TOPSIS for group decision making under fuzzy environment. Fuzzy Sets Syst. **114**, 1–9 (2000)
19. Chen, C.T., Lin, C.T., Huang, S.F.: A fuzzy approach for supplier evaluation and selection in supply chain management. Int. J. Prod. Econ. **102**, 289–301 (2006)
20. Zadeh, L.A.: Soft computing and fuzzy logic. In IEEE Softw. **11**(6), 48–56 (1994)
21. Sadiq, M.: A fuzzy set-based approach for the prioritization of stakeholders on the basis of the importance of software requirements. IETE J. Res. **63**(5), 1–14 (2017)

A Fuzzy-Based Multi-Criteria Decision-Making Approach for the Selection of Digital Image Forensic Tools

Azra Parveen, Zishan Husain Khan, and Syed Naseem Ahmad

Abstract Digital image forensic science is a sub-research area of multimedia security whose objective is to check the authenticity of digital images. Different algorithms as well as tools have been developed to check the forged images. In literature, less attention is given on the evaluation and selection of the digital image forensic tools based on different features like error level analysis, metadata analysis, double joint photographic expert group, etc. Therefore, to address this issue, in this chapter an algorithm has been developed for the selection of the digital image forensic tools based on the ranking values. The ranking values of the digital image forensic tools are computed using TOPSIS method by using the triangular fuzzy numbers. The utilization of the proposed method is discussed with the help of an example in which following tools have been considered during the analysis, i.e., FotoForensics, JPEGsnoop, Forensically, Ghiro, and Izitru.

Keywords Digital image forensic · Multi-criteria decision-making · Fuzzy logic · Technique for order of preference by similarity to ideal solutions · TOPSIS

Abbreviations

AHP	Analytic Hierarchy Process
AIFD	Active Image Forgery Detection
C	Criteria
CC	Closeness Coefficients

A. Parveen (✉) · Z. H. Khan
Department of Applied Sciences and Humanities, Jamia Millia Islamia (A Central University), New Delhi 110025, India

Z. H. Khan
e-mail: zishanhk@jmi.ac.in

S. N. Ahmad
Department of Electronics and Communication Engineering, Jamia Millia Islamia (A Central University), New Delhi 110025, India
e-mail: snahmad@jmi.ac.in

A. J. Kulkarni (ed.), *Multiple Criteria Decision Making*, Studies in Systems, Decision and Control 407, https://doi.org/10.1007/978-981-16-7414-3_5

DIF	Digital Image Forensic
DM	Decision-Makers
ELA	Error Level Analysis
FDM	Fuzzy Decision Matrix
FNIS	Fuzzy Negative Ideal Solution
FPIS	Fuzzy Positive Ideal Solution
FST	Fuzzy Set Theory
H	High
JPEG	Joint Photographic Expert Group
L	Low
M	Medium
MA	Metadata
MCDM	Multi-Criteria Decision-Making
PIFD	Passive Image Forgery Detection
S	Strong
T	Tools
TFNs	Triangular Fuzzy Numbers
TOPSIS	Technique for Order of Preference by Similarity to Ideal Solution
VH	Very High
VL	Very Low
VS	Very Strong
VW	Very Weak
W	Weak
WN	Weighted Normalized Fuzzy Decision Matrix

List of Symbols

$\sum$	Summation
min ()	Returns minimum value
max ()	Returns maximum value

1 Introduction

In computer and electronic science, security is a non-functional requirement whose objective is to safeguard the valuable data or information from unlawful users. Research in the area of security has been divided into following: information security, data security and privacy, cloud computing security, multimedia security, network security, and Internet of Things security [1]. In this chapter, we mainly focused on one of the research areas of the multimedia security, i.e., digital image forgery. With the development of the advanced tools like Photoshop, Corel Draw, etc., it is easy

(a) Original image (b) Forged image

Fig. 1 Original and forged image of a child [3]

to change the contents of an image. As a result, it produces the fake images. It is important to identify the fake images from the original images. The identification of the fake images from the set of the images is not an easy task for the naked eyes. Therefore, the objective of the image forensic science research is to develop the algorithms and tools for the detection of the forged images [1, 2]. The issues related with the images have been started since 1860. In the database of the image authentication service curated by fourandsix technologies, there are 245 forged images [3]. From this database, it was found that the original image of a chubby girl who is drinking milk was forged by a bag labeled sugar above a caption that warned that the sugary drinks can cause obesity, as shown in Fig. 1.

Different algorithms and tools have been developed for the detection of the forged images. For example, Parveen et al. [1] proposed a method for the detection of the forged images using discrete cosine transform. In another study, Parveen and Tayal [4] developed an algorithm for the detection of the forged images using color filter array. In addition to these algorithms, different tools have also been developed for the detection of the doctored images. For example, FotoForensics tool, JPEGsnoop tool, Ghiro tool, Forensically tool, and Izitru tool. In our recent study [5], these tools have been evaluated on the basis of the following features: (i) error level analysis, (ii) metadata analysis, (iii) last save quality, (iv) JPEG luminance and chrominance, (v) digest, (vi) file type extension and MIME type, (vii) image width and height, (viii) bits per sample, (ix) color components, (x) cryptographic hash function, (xi) clone detection, (xii) principal component analysis, (xiii) noise analysis, (xiv) GPS-Localization, (xv) devise signature analysis, (xvi) double JPEG detection, (xvii) JPEG structure/coefficients/ghost detection, and (xviii) sensor pattern analysis. The objective of this chapter is to extend our previous work [5] using multi-criteria decision-making (MCDM) method for the selection of digital image forensic (DIF) tools based on the features.

Selection of DIF tools based on different features is an MCDM problem whose objective is to select the DIF tools for the detection of the doctored images. Due to the increasing complexity of the socio-economic environment, several decision-makers are involved during the DIF tools because it is difficult for the single decision-maker to deal with all the features of the tools [6, 7]. During the decision-making process, decision-makers may use linguistic variables to specify their preferences for the evaluation of the tools based on different features. For example, the tool should support "*more*" on error level analysis (ELA). Here, the term *more* is a linguistic variable. There are vagueness and imprecision in human judgement. Different mathematical tools have been developed to deal with vagueness and impression during the decision-making process like fuzzy set theory, rough set theory, etc. In literature, different MCDM methods have been developed like "Analytic Hierarchy Process" (AHP), "Technique for Order of Preference by Similarity to Ideal Solution" (TOPSIS), etc. In this chapter, fuzzy TOPSIS has been used for the selection of DIF tools based on the features because it is difficult to deal with vagueness and imprecision using exact numbers [8, 9].

The remaining part of this chapter is organized as follows: Related work is discussed in Sect. 2. An insight into fuzzy set theory is given in Sect. 3. An evaluation of DIF tools based on different features is discussed in Sect. 4. An example is given in Sect. 5. Finally, Sect. 6 concludes the chapter.

2 Related Work

One of the key research areas of the operation research is the MCDM methods whose objective is to develop the computational tools for the subjective evaluation based on the different criteria. The MCDM methods have been used for evaluating, accessing, and ranking alternatives in the following areas like (a) "information technology and systems", (b) "supply chain management", (c) "business and marketing management", (d) "design engineering and manufacturing", etc. [10].

In the area of the digital image forensic science, different methods have been developed to detect the image forgery, and these methods are broadly divided into two parts, i.e., "active image forgery detection (AIFD) methods" and "passive image forgery detection (PIFD) methods". In AIFD methods, "a watermark is embedded in the image". To examine the genuineness of the image, the embedded watermark is retrieved from the image. If the "extracted watermark" is same as the "*original watermark*", then the image is considered as the genuine image, else, it is treated as the doctored image. In practical situations, we don't have the prior information about the watermark; therefore, we do not pay more attention on AIFD methods. In image forgery research area, most of the work is dedicated to the PIFD methods in which no prior information about the images are required to check whether the images are forged or not [11, 12]. In addition to these techniques, there are also image forgery detection tools which are used in real-life application for the detection of the image forgery. The evaluation of these tools and their analysis is discussed in our previous work [5, 13]. PIFD methods are classified into five sub-parts, i.e., "*pixel based*", "*compression based*", "*camera based*", "*physics based*", and "*geometric based*" techniques [1, 11, 12].

To identify the research gaps in the literature, we have performed a systematic literature review in the area of pixel-based image forgery detection techniques [14] using the guidelines of Kitchenham [15]. Based on our review, we found that most of the focus in image forgery detection techniques is given to different types of image feature extraction methods, i.e., (1) Discrete Cosine Transform; (2) Discrete Wavelet Transform; (3) Principal Component Analysis; (4) Signal Value Decomposition; (5) Histogram of Oriented Gradients; (6) Zernike Moment; (7) Fourier Mellin Transform; (8) Polar Complex Exponential Transform; (9) Fourier Transform; (10) Polar Cosine Transform; (11) Patch-Match algorithm; (12) Polar Harmonic Transform; (13) Local Binary Patterns; (14) Blur Invariant Moment; (15) Polar Coordinate System; (16) Scale Invariant Feature Transform; (17) Speedup Robust Features; (18) J-Linkage algorithm; (19) Harris Corner Points. In literature, the following methods have been successfully applied for feature matching in image forgery detection, i.e., (1) Exhaustive search, (2) Lexicographically sorting, (3) K-Dimensional Tree, (4) Radix Sorting, (5) Counting Bloom Filters, (6) Best-Bin-First [1]. Based on our review, we found that in literature less attention is given to the selection of the DIF tools based on the features using MCDM techniques. Therefore, it motivates us to work in the DIF tools and apply fuzzy TOPSIS for the selection of the DIF tools based on features.

3 Fuzzy Set Theory

Fuzzy set theory (FST) is an important component of soft computing which is used to deal with the imprecision and vagueness during the decision-making process. The notion of fuzzy logic was developed by Lotfi A. Zadeh in 1965 to deal with the linguistic variables. FST is a multi-valued logic; on the other hand, crisp set theory is Boolean logic. In crisp logic, the elements are either present or not in a set; and in this theory there is no concept of partial membership of the elements. In FST, the partial membership values of the elements are also considered during the decision-making process. Consider the following fuzzy set $Z = \left\{\frac{0.6}{x1} + \frac{0.8}{x2} + \frac{0.2}{x3}\right\}$; in this set, the degree of membership values of $x1$, $x2$, and $x3$ are 0.6, 0.8, and 0.2, respectively [16].

In FST, linguistic variables are modeled by different membership functions like triangular membership function, trapezoidal membership function, bell-shaped membership function, etc. These membership functions are represented by fuzzy numbers like triangular fuzzy numbers (TFNs), etc. For example, if the linguistic variable "Very Strong" (VS) is represented by $(0.7, 0.8, 0.9)$. In this example, the linguistic variable VS is modeled by a TFN in which the value 0.9 is optimistic estimate, "which is intended to be the unlikely but possible value if everything goes well"; the value 0.8 is the most likely estimate, "intended to be the most realistic value"; and the value 0.7 is a pessimistic estimate, "which is intended to be the

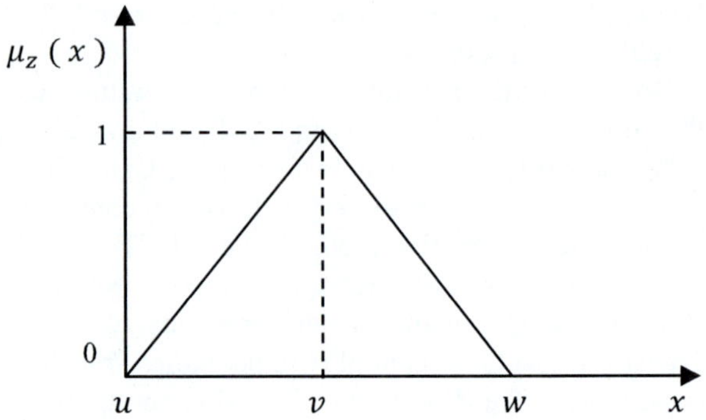

Fig. 2 The membership function of a TFN $Z = (u, v, w)$

unlikely but possible value if everything goes badly". There are different applications of FST in the area of science and engineering. For example, medical sciences, wireless sensor networks, software engineering, management science, etc. [17, 18]. Among various fuzzy numbers, in this chapter, TFNs have been used because of its simplicity in understanding and computation and it is represented by $Z = (u, v, w)$, as shown in Fig. 2.

$$\mu_T(x) = \begin{cases} 0 & x \leq u \\ \frac{x-u}{v-u} & u \leq x \leq v \\ \frac{w-x}{w-v} & v \leq x \leq w \\ 0 & w \leq x \end{cases} \tag{1}$$

There are different operations that can be performed on TFNs like sum, difference, inverse, etc. Suppose $Z_1 = (u_1, v_1, w_1)$ and $T_2 = (u_2, v_2, w_2)$ are two TFNs, then:

$$(u_1, v_1, w_1) + (u_2, v_2, w_2) = (u_1 + u_2, v_1 + v_2, w_1 + w_2) \tag{2}$$

$$(u_1, v_1, w_1) . (u_2, v_2, w_2) = (u_1 . u_2, v_1.v_2, w_1.w_2) \tag{3}$$

$$(u_1, v_1, w_1)^{-1} \approx (\frac{1}{w_1}, \frac{1}{v_1}, \frac{1}{u_1}) \tag{4}$$

$$(u_1, v_1, w_1) . k = (u_1 k, v_1 k, w_1 k) \tag{5}$$

where k is a positive real number.

The distance between two TFNs can be computed by vertex method [7, 18].

$$\boldsymbol{d}(Z_1, Z_2) = \sqrt{\frac{1}{3}\left[(u_1 - u_2)^2 + (v_1 - v_2)^2 + (w_1 - w_2)^2\right]} \tag{6}$$

4 A Fuzzy TOPSIS Method for the Selection of Digital Image Forensic Tools

The objective of this section is to present the proposed fuzzy TOPSIS method for the selection of the DIF tools based on features. The block diagram of the proposed method is exhibited in Fig. 3. There are six steps in the proposed method, i.e., (i) identification of the DIF tools, (ii) define decision-maker's linguistic variables, (iii) construct the fuzzy decision matrix for digital image forensic tool, (iv) construct the normalized and weighted normalized decision matrix, (v) compute fuzzy positive and fuzzy negative ideal solutions, and (vi) calculate the closeness coefficients of each DIF tool. The explanation of these steps is given as below:

Step 1: Identification of the DIF tools and its features

The objective of this step is to identify those tools that will be evaluated based on the features or criteria. There are different DIF tools in the literature which are used for the detection of the forged images. In our work, traditional method of the software requirements elicitation process has been used for the identification of the tools and its features. Traditional method is sub-divided into (a) interview, (b) analysis of the existing documents, and (c) questionnaire [19].

Step 2: Define decision-maker's linguistic variables

Before defining the linguistic variables of the decision-makers, it is necessary to form the committee of the decision-makers (DM) who will participate in the selection of the DIF tools. Here, it is assumed that, M decision-makers are participating in DIF tools selection process; and the fuzzy rating of the DMs on DIF tools and features are

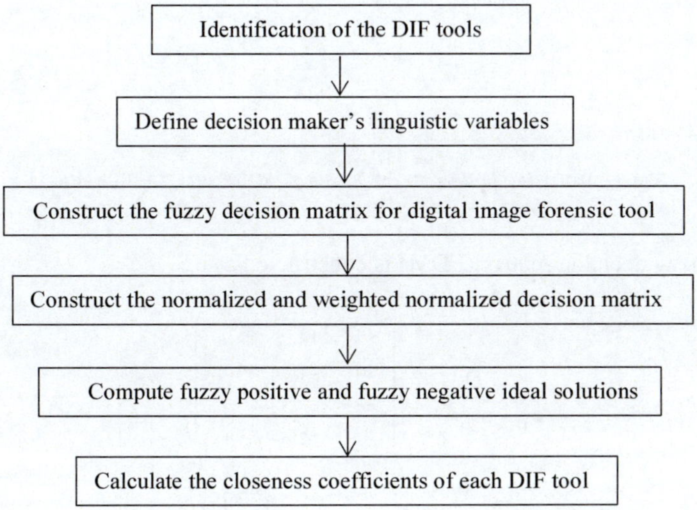

Fig. 3 Block diagram of proposed method

represented by TFNs, $Z_i = (u_i, v_i, w_i), i = 1, 2, \ldots, I$. In real-life applications, it has been observed that DMs use linguistic variables instead of crisp numbers, therefore, the objective of this step is to define the linguistic variables that will be used during the decision-making process.

Step 3: Construct the fuzzy decision matrix for digital image forensic tool

In this step, the fuzzy decision matrix (FDM) is formed after the evaluation of the DIF tools based on the features by M decision makers. The fuzzy ratings of the DMs are defined as $Z_i = (u_i, v_i, w_i), i = 1, 2, \ldots, I$. The aggregated fuzzy rating can be computed as $Z = (u, v, w), i = 1, 2, .., I$ [20]. Here,

$$u = \min\{u_i\}, v = \frac{1}{I}\sum_{i=1}^{I} v_i, \ and\ w = \max\{w_i\} \tag{7}$$

Suppose the fuzzy rating of the ith DM is $x_{abi} = (u_{abi}, v_{abi}, w_{abi})$, where $a = 1, 2, \ldots, p$, and $b = 1, 2, \ldots, q$. Here, p and q are the number of DIF tools and features, respectively. The aggregated fuzzy rating (x_{ab}) of DIF tools with respect to features can be calculated as $(x_{ab}) = (u_{ab}, v_{ab}, w_{ab})$. Here,

$$\begin{aligned} u_{ab} &= \boldsymbol{min}\{u_{abi}\},\ v_{ab} = \frac{1}{I}\sum_{i=1}^{I} v_{abi}\ and \\ w_{ab} &= \max\{w_{abi}\}, \end{aligned} \tag{8}$$

Suppose the importance weight of the ith DMs is $w_{abi} = (w_{bi1}, w_{bi2}, w_{bi3})$, then the aggregated fuzzy weights (w_{ab}) of each feature is computed as:

$$w_b = (w_{b1}, w_{b2}, w_{b3})$$

Here,

$$w_{b1} = \min\{u_{bi1}\}, w_{b2} = \frac{1}{I}\sum_{i=1}^{I} v_{bi2},\ and\ w_{b3} = \max\{w_{bi3}\} \tag{9}$$

The fuzzy decision matrix (FDM) is constructed as:

$$FDM = \begin{bmatrix} x_{11} & x_{12} & \ldots & x_{1q} \\ x_{21} & x_{22} & \ldots & x_{2q} \\ \ldots & \ldots & \ldots & \ldots \\ x_{p1} & x_{p2} & \ldots & x_{pq} \end{bmatrix}$$

$$W = \left[w_1, w_2, \ldots, w_q\right]$$

Here, $x_{ab} = (u_{ab}, v_{ab}, w_{ab})$ and $w_b = (w_{b1}, w_{b2}, w_{b3})$; $a = 1, 2, \ldots, p, b = 1, 2, \ldots, q$ can be approximated by positive TFNs.

Step 4: Construct the normalized and weighted normalized decision matrix

In this step, the normalized FDM is obtained by using the linear scale transform (LST):

$$LST = [t_{ab}]_{p\times q}, a = 1, 2, \ldots, p \; and \; b = 1, 2, \ldots, q \tag{10}$$

where

$$t_{ab} = \left\{\frac{u_{ab}}{w_b^*}, \frac{v_{ab}}{w_b^*}, \frac{w_{ab}}{w_b^*}\right\}$$

$$w_b^* = max \; w_{ab}$$

The weighted normalized FDM is computed by multiplying the importance weights of each feature of the DIF tool and the values of the normalized FDM. The weighted normalized FDM (WN) is defined as:

$$WN = [wn_{ab}]_{p\times q}, a = 1, 2, \ldots, p \; and \; b = 1, 2, \ldots, q \tag{11}$$

$$wn_{ab} = t_{ab}(.)w_b$$

Here, w_b represents the importance weight of each feature of the DIF tool.

Step 5: Compute fuzzy positive and fuzzy negative ideal solutions

The objective of this step is to compute the fuzzy positive ideal solutions (FPIS) and fuzzy negative ideal solutions (FNIS) as:

$$FPIS = (FPIS_1, FPIS_2, \ldots, FPIS_q) \tag{12}$$

$$FNIS = (FNIS_1, FNIS_2, \ldots, FNIS_q) \tag{13}$$

where

$$FPIS_b = \max\{FPIS_{ab3}\} \; and \; FNIS_b = \min\{FNIS_{ab1}\}$$

$$a = 1, 2, \ldots, p \; and \; b = 1, 2, \ldots, q$$

Step 6: Calculate the closeness coefficients of each DIF tool

The closeness coefficient (CF) of each digital image forensic tool is computed as:

$$CF_a = \frac{d_a^{FNIS}}{d_a^{FPIS} + d_a^{FNIS}}, a = 1, 2, \ldots, p \quad (14)$$

where

$$d_a^{FPIS} = \sum_{b=1}^{q} d(wn_{ab}, FPIS_b) \; a = 1, 2, \ldots, p \quad (15)$$

$$d_a^{FNIS} = \sum_{b=1}^{q} d(wn_{ab}, FNIS_b) \; a = 1, 2, \ldots, p \quad (16)$$

where $d(.,.)$ is the distance between two fuzzy numbers.

5 An Example

To explain the steps of the proposed method, we have considered the DIF tools which are used for the detection of the doctored images. Based on our literature review [5], we have identified the following digital image forensic tools (T), i.e., T1: FotoForensics, T2: JPEGsnoop, T3: Forensically, T4: Ghiro, and T5: Izitru, which are used for the identification of the forged portion in digital images. In this chapter, following criteria (C) have been used for the evaluation of the digital image forensic tools, i.e., C1: "error level analysis" (ELA), C2: "metadata" (MA), and C3: "double joint photographic expert group" (D-JPEG). In this chapter, the following linguistic variables have been used for the evaluation of the features of the DIF tools, i.e., Very Weak (VW) = (2, 2, 4), Weak (W) = (2, 4, 6), Medium (M) = (4, 6, 8), Strong (S) = (6, 8, 10), and Very Strong (VS) = (8, 10, 10). For the evaluation of the relationship between DIF tools and criteria, the following linguistic variables have been used: Very Low = (0.0, 0.0, 0.25), Low (L) = (0.0, 0.25, 0.5), Medium (M) = (0.25, 0.5, 0.75), High (H) = (0.5, 0.75, 1.0), and Very High (VH) = (0.75, 1.0, 1.0). In this chapter, it is assumed that three decision-makers are participating during the selection process of the DIF tools.

To construct the fuzzy decision matrix (FDM), the digital image forensic tools are first evaluated by the three decision-makers (DM), i.e., DM_1, DM_2, and DM_3; and the weight of the criteria's are also evaluated based on the linguistic variables. The results after the evaluation of digital forensic tools and the criteria are exhibited in Tables 1 and 2.

Here, the weighted FDM is constructed by using Eqs. (10) and (11). The weighted FDM is exhibited in Table 3.

Fuzzy positive ideal solution (FPIS) and fuzzy negative ideal solution (FNIS) are computed with the help of the Eqs. (12) and (13).

Table 1 Evaluation of digital image forensic tools under three criteria

Criteria	Digital image forensic tools	Decision-makers		
		DM_1	DM_2	DM_3
C1	T1	VH	H	VH
	T2	M	VL	L
	T3	VH	M	VL
	T4	VH	H	VH
	T5	VL	VL	L‘
C2	T1	VH	VH	H
	T2	VH	VH	H
	T3	VL	L	L
	T4	VH	VH	H
	T5	VL	L	L
C3	T1	VL	L	VL
	T2	VL	L	L
	T3	VL	VL	VL
	T4	VH	H	L
	T5	VH	VH	H

$$FPIS = [(1.0, 1.0, 1.0), (1.0, 1.0, 1.0), (1.0, 1.0, 1.0)]$$

$$FNIS = [(0.0, 0.0, 0.0), (0.0, 0.0, 0.0), (0.0, 0.0, 0.0)]$$

The distance of each DIF tool from $FPIS$ and $FNIS$ with respect to each criterion is calculated by using the Eq. (6); and the results are exhibited in Tables 4 and 5.

$$d(T1, FPIS) = \sqrt{\frac{(1 - 0.2)^2 + (1 - 0.67)^2 + (1 - 1)^2}{3}} = 0.4996$$

$$d(T1, FNIS) = \sqrt{\frac{(0 - 0.2)^2 + (0 - 0.67)^2 + (0 - 1)^2}{3}} = 0.7045$$

Equation (14) is used to compute the closeness coefficients of each DIF tool. The closeness coefficient for the DIF tools is given as below:

$$\text{Closeness coefficient of T1} = \frac{1.7235}{1.8249 + 1.7235} = 0.4857$$

$$\text{Closeness coefficient of T2} = \frac{1.1632}{2.057 + 1.1632} = 0.3612$$

Table 2 Fuzzy decision matrix and fuzzy weights of three criteria

Criteria/tools	T1	T2	T3	T4	T5
C1 (0.4, 0.733, 1.0)	(0.5, 0.92, 1.0)	(0, 0.25, 0.75)	(0.0, 0.5, 1.0)	(0.5, 0.92, 1)	(0.0, 0.08, 0.5)
C2 (0.4, 0.8, 1.0)	(0.5, 0.92, 1.0)	(0.5, 0.92, 1.0)	(0.0, 0.167, 0.5)	(0.5, 0.92, 1)	(0, 0.167, 0.5)
C3 (0.6, 0.87, 1.0)	(0.0, 0.08, 0.5)	(0.0, 0.167, 0.5)	(0.0, 0.0, 0.25)	(0.0, 0.67, 1)	(0.5, 0.92, 1)

Table 3 Weighted fuzzy decision matrix

Criteria/tools	T1	T2	T3	T4	T5
C1	(0.2, 0.67, 1.0)	(0.0, 0.18, 0.75)	(0.0, 0.37, 1.0)	0.2, 0.67, 1.0)	(0.0, 0.05, 0.5
C2	(0.2, 0.74, 1.0)	(0.2, 0.74, 1.0)	(0.0, 0.13, 0.5)	(0.2, 0.74, 1.0)	(0.0, 0.13, 0.5)
C3	(0.0, 0.07, 0.5)	(0.0, 0.15, 0.5)	(0.0, 0.0, 0.25)	(0.0, 0.58, 1.0)	(0.3, 0.80, 1.0)

Table 4 The distance between digital image forensic tools T_i $(i = 1, 2, \ldots, 5)$ and $FPIS$ with respect to three criteria

Distance between Ts and $FPIS$	T1	T2	T3	T4	T5
C1	0.4996	0.7604	0.6824	0.4996	0.8470
C2	0.4857	0.4857	0.8179	0.4857	0.4857
C3	0.8396	0.8109	0.9242	0.6262	0.4203
Sum	1.8249	2.057	2.4245	1.6115	1.753

Table 5 The distance between digital image forensic tools T_i $(i = 1, 2, \ldots, 5)$ and $FNIS$ with respect to three criteria

Distance between Ts and $FNIS$	T1	T2	T3	T4	T5
C1	0.7045	0.1343	0.6156	0.704	0.2901
C2	0.7275	0.7275	0.2983	0.7275	0.2983
C3	0.2915	0.3014	0.1443	0.6674	0.7594
Sum	1.7235	1.1632	1.0582	2.0989	1.3478

$$\text{Closeness coefficient of T3} = \frac{1.0582}{2.4245 + 1.0582} = 0.3038$$

$$\text{Closeness coefficient of T4} = \frac{2.0989}{1.6115 + 2.0989} = 0.5657$$

$$\text{Closeness coefficient of T5} = \frac{1.3478}{1.753 + 1.3478} = 0.4347$$

Based on the closeness coefficient value of the DIF tools, we found that DIF tool T4 has the highest priority and T3 has the lowest priority. Therefore, based on the evaluation of the DIF tools, the Ghiro tool (T4) will be used for the detection of the doctored images because it has the highest priority.

6 Conclusion and Future Directions

In this chapter, we have presented a method for the evaluation and selection of the DIF tools using fuzzy TOPSIS. The proposed method includes six steps, i.e., (i) identification of the DIF tools, (ii) define decision-makers linguistic variables, (iii) construct the fuzzy decision matrix for DIF tool, (iv) construct the normalized and weighted normalized decision matrix, (v) compute fuzzy positive and fuzzy negative ideal solutions, and (vi) calculate the closeness coefficients of each DIF tool. In this chapter, TFNs were used to model the linguistic variables during the evaluation process. The proposed method has been applied for the selection of the five DIF tools based on three criteria. Based on the closeness coefficient, it was found that Ghiro tool has the highest priority value. In future, we will try to work on the following:

- To develop a tool for the selection of the DIF tools
- To apply fuzzy AHP for the evaluation and selection of the DIF tools

References

1. Parveen, A., Khan, Z.H., Ahmad, S.N.: Block-based copy-move image forgery detection using DCT. Iran J. Comput. Sci. **2**, 89–99 (2019)
2. Qureshi, M.A., Deriche, M.: A bibliography of pixel-based blind image forgery detection techniques. Signal Process. Image Commun. **39**(A), 46–74 (2015)
3. Photo Tampering Throughout History. http://pth.izitru.com/1860_13_00.html
4. Parveen, A., Tayal, A.: An algorithm to detect the forged part in an image. In: IEEE International Conference on Communication and Signal Processing, pp. 1486–1490 (2016)
5. Parveen, A., Khan, Z.H., Ahmad, S.N.: Classification and evaluation of digital forensic tools. TELKOMNIKA Telecommun. Comput. Electron. Control **18**(6), 3096–3106 (2020)
6. Mardani, A., Jusoh, A., Nor, K. MD., Khalifah, Z., Zakwan, N., Valipour, A.: multiple criteria decision making techniques and their applications-a review of the literature from 2000 to 2014. Econ. Res. **28**(1), 516–571 (2015)
7. Sadiq, M., Khan, S., Mohammad, C.W.: Selection of software requirements using TOPSIS under fuzzy environment. Int. J. Comput. Appl. 1–10 (2020)
8. Chen, C.T.: Extension of the TOPSIS for group decision making under fuzzy environment. Fuzzy Sets Syst. **114**, 1–9 (2000)
9. Ertugrul, I., Karakasoglu, N.: Comparison of fuzzy AHP and fuzzy TOPSIS methods for facility location selection. Int. J. Adv. Manuf. Technol. **39**, 783–795 (2008)
10. Behzadian, M., Otaghsara, S.K., Yazdani, M., Ignatius, J.: A State-of the-art survey of TOPSIS applications. Expert Syst. Appl. **39**, 13051–13069 (2012)
11. Kumar, M., Srivastava, S.: Image forgery detection based on physics and pixels. Aust. J. Forensic Sci. **51**(2), 119–134 (2019)
12. Mahdian, B., Saic, S.: A bibliography on blind methods for identifying image forgery. Signal Proces. Image Commun. **25**(6), 389–399 (2010)
13. Parveen, A., Khan, Z.H., Ahmad, S.N.: Identification of the forged images using image forensic tools. In: Proceedings of 2nd International Conference on Communication and Computing Systems, CRC-Press, Taylor and Francis, pp. 1–6 (2018)
14. Parveen, A., Khan, Z.H., Ahmad, S.N.: Pixel based copy-move image forgery detection techniques: a systematic literature review. In: Proceedings of the 5th IEEE International Conference

on Computing for Sustainable Global Development, organized by BVICAM, New Delhi, India, pp. 663–668, (2018).

15. Kitchenham, B.: Procedure for performing Systematic Review, Joint Technical Report, Software Engineering Group, Department of Computer Science. Keele University, United Kingdom and Empirical Software Engineering, national ICT Australia (2004)
16. Zadeh, L.A.: Fuzzy logic = computing with words. IEEE Trans. Fuzzy Syst. **4**(2), 103–111 (1996)
17. Sadiq, M., Jain, S.K.: Applying fuzzy preference relation for requirements prioritization in goal oriented requirements elicitation process. Int. J. Syst. Assur. Eng. Manag. **5**(4), 711–723 (2014)
18. Shih, H.-S., Shyur, H.-J., Lee, E.S.: An extension of TOPSIS for group decision making. Math. Comput. Modell. **45**, 801–813 (2007)
19. Sadiq, M., Jain, S.K.: An insight into requirements engineering processes. In: 3rd International Conference on Advances in Communication, Network, and Computing LNCSIT, Chennai, pp. 313–318 (2012)
20. Chen, C.T.: Extensions of the TOPSIS for group decisions making under fuzzy environment. Fuzzy Sets Syst. **114**, 1–9 (2000)

Why Does the Choice of Normalization Technique Matter in Decision-Making

Andrii Shekhovtsov, Aleksandra Kaczyńska, and Wojciech Sałabun

Abstract Multi-Criteria Decision Analysis (MCDA) methods are very important to help the decision-maker to make more responsible choices. Despite creating new techniques and improving existing ones, each decision problem has a set of criteria and alternatives that are presented in a decision matrix. Most MCDA methods require normalization of this matrix due to different units of measurement that are not suitable for direct comparison. However, it should be noted that any normalization can contribute to change in the final result. In this chapter, we present a simple investigation to show the fundamental differences between the five most common normalization techniques. We used these methods on randomly generated diverse data sets and carried out a comparison of necessary statistical data. It turned out that the characteristics of data sets have a significant impact on normalization results.

Keywords Normalization · MCDA · Decision support

Abbreviations

MCDA	Multi-Criteria Decision Analysis.
TOPSIS	Technique for Order Preference by Similarity to Ideal Solution.
PROMETHEE	Preference Ranking Organization Method for Enrichment of Evaluations.
VIKOR	VlseKriterijumska Optimizacija I Kompromisno Resenje.
COPRAS	Complex Proportional Assessment.

A. Shekhovtsov · A. Kaczyńska · W. Sałabun (✉)
West Pomeranian University of Technology in Szczecin, ul. Żołnierska 49, 71-210 Szczecin, Poland
e-mail: wojciech-salabun@zut.edu.pl

A. Shekhovtsov
e-mail: andrii-shekhovtsov@zut.edu.pl

A. Kaczyńska
e-mail: aleksandra-kaczynska@zut.edu.pl

A. J. Kulkarni (ed.), *Multiple Criteria Decision Making*, Studies in Systems, Decision and Control 407, https://doi.org/10.1007/978-981-16-7414-3_6

Mathematical Symbols

$X_{n \times m}$ Decision matrix, where n stands for the number of alternatives and m for the number of criteria.

x_{ij} Element of the decision matrix, where $i = 1, 2, \ldots, n$ is an index of alternative and $j = 1, 2, \ldots, m$ is an index of criteria.

r_{ij} Element of the normalized decision matrix, where $i = 1, 2, \ldots, n$ is an index of alternative and $j = 1, 2, \ldots, m$ is an index of criteria.

1 Introduction

Many criteria, including those that are incompatible with each other, often need to be considered when making decisions. MCDA (Multi-Criteria Decision-Analysis) methods are an important tool when dealing with such situations. They are widely used in many areas to help the decision-maker to find a solution that best suits his expectations.

In most decision-making problems, it is necessary to have properly prepared data. An essential part of the process is to create a decision matrix in which the data must be normalized [3, 15]. This situation occurs in many known MCDA methods, which cannot work correctly without preliminary data normalization, the best example being TOPSIS [8, 13, 14]. It is a very popular approach because of its simplicity, but at the same time, it is susceptible to choose the normalization method. Of course, some methods can work without normalization such as PROMETHEE and VIKOR [7, 12]. However, most often, this normalization is still desired in order to achieve better comparability of input data [4]. Sometimes the type of normalization is indicated from above as in the case of Entropy or COPRAS methods [10, 16, 17]. However, in this case, there are modifications, which often consist of changing the current normalization method.

This issue leads to the question of which normalization method to choose [1, 2, 12]. It depends on the specificity of a given problem, i.e., on the characteristics of the input data [3]. In this chapter, we present a short comparison of the five most commonly used techniques in six different scenarios. We aim to show what happens to the data after the normalization in the selected settings. Very often in publications, the authors do not refer at all to the justification of the selection of normalization technique. It looks as if it is a random procedure, and yet it determines the final result [6, 9, 13].

In the conducted experiments, six sets of data were randomly generated, differing in range, size, and sign. Then, if possible, the following normalization methods were applied to them: minimum–maximum, max, sum, vector, and logarithmic. The study was omitted for sets containing negative numbers in case of logarithmic normalization. Then, for each set of data, there was a comparison of statistics (minimum, maximum, average, quantiles) of the original set and the results obtained with the tested methods.

2 Normalization Methods

In the literature, there is no clear assignment to which decision-makers' methods of data normalization are used. This situation poses a problem, as it is necessary to consider the influence of particular normalization on the result. The most common normalization methods in MCDA methods can be divided into two groups [5, 11, 13], i.e., methods designed to profit (1), (3), (5), (7) and cost criteria (2), (4), (6), (8).

The minimum–maximum method—in this approach, the greatest and the least values in the considered set are used. The formulas are described as follows (refer Eqs. 1 and 2):

$$r_{ij} = \frac{x_{ij} - min_j(x_{ij})}{max_j(x_{ij}) - min_j(x_{ij})}, \tag{1}$$

$$r_{ij} = \frac{max_j(x_{ij}) - x_{ij}}{max_j(x_{ij}) - min_j(x_{ij})}. \tag{2}$$

The maximum method—in this technique, only the greatest value in the considered set is used. The formulas are described as follows (refer Eqs. 3 and 4):

$$r_{ij} = \frac{x_{ij}}{max_j(x_{ij})}, \tag{3}$$

$$r_{ij} = 1 - \frac{x_{ij}}{max_j(x_{ij})}. \tag{4}$$

The sum method—in this method, the sum of all values in the considered set is used. The formulas are described as follows (refer Eqs. 5 and 6):

$$r_{ij} = \frac{x_{ij}}{\sum_{i=1}^{m} x_{ij}}, \tag{5}$$

$$r_{ij} = \frac{\frac{1}{x_{ij}}}{\sum_{i=1}^{m} \frac{1}{x_{ij}}}. \tag{6}$$

The vector method—in this method, the square root of the sum of all values. The formulas are described as follows (refer Eqs. 7 and 8):

$$r_{ij} = \frac{x_{ij}}{\sqrt{\sum_{i=1}^{m} x_{ij}^2}}, \tag{7}$$

$$r_{ij} = 1 - \frac{x_{ij}}{\sqrt{\sum_{i=1}^{m} x_{ij}^2}}. \tag{8}$$

The logarithmic method—in this method of normalization uses the natural logarithm. Values of considered set are assumed to be positive. The formulas are described for profit type (refer Eq. 9) and cost type (refer Eq. 10) as follows:

$$r_{ij} = \frac{\ln\left(x_{ij}\right)}{\ln\left(\prod_{i=1}^{m} x_{ij}\right)}, \tag{9}$$

$$r_{ij} = \frac{1 - \frac{\ln(\Delta_j)}{\ln(I\,\lim_{i=1} A_j)}}{m - 1}. \tag{10}$$

3 Experiments

3.1 Set with Natural Numbers

The first set of data contain ten consecutive natural values from one to ten. Figure 1 shows the data set before and after normalization.

The minimum–maximum and maximum methods very adequately reproduce the shape of the set, which is very similar to raw data. In the case of other methods, the results grow slower than in the original. The sum method gives results similar in shape to the logarithmic function because, in both cases, the results are the flattest.

The box plot (Fig. 2) shows that data normalized by the minimum–maximum method (1) is evenly distributed over the range from 0 to 1. The maximum method (2) gives similar results, including a similar average, but the minimum value is equal

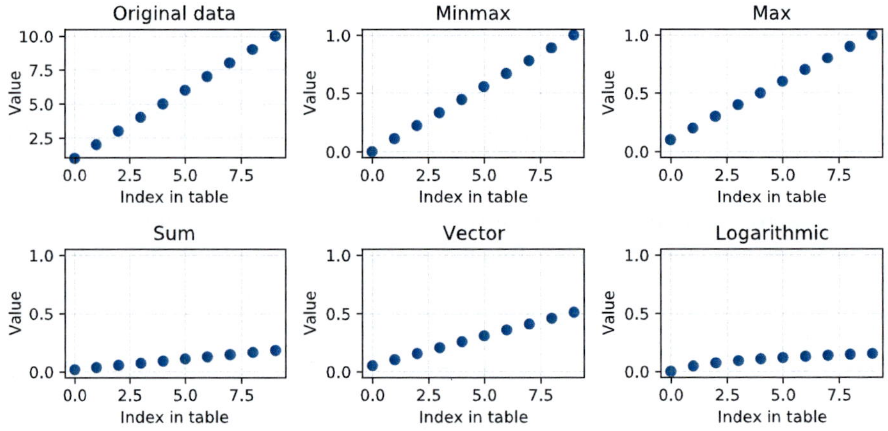

Fig. 1 Visualization of data set Sect. 3.1 before and after normalization

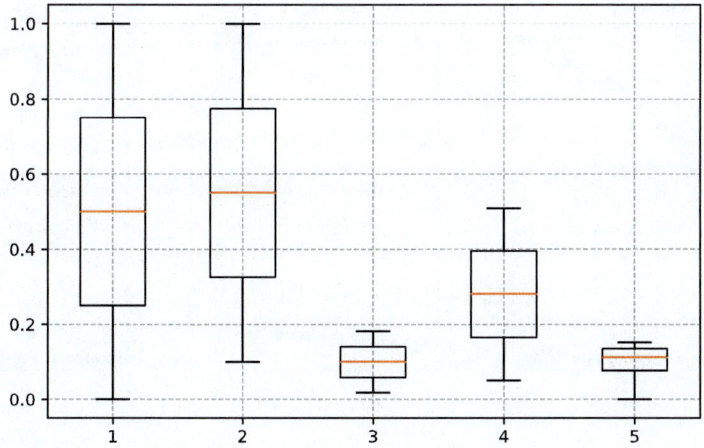

Fig. 2 Box plot for data set Sect. 3.1, where 1—minmax, 2—max, 3—sum, 4—vector, and 5—logarithmic

Table 1 Statement of statistical parameters for data set Sect. 3.1

Normalization method	Min	1st Qu.	Median	Mean	3rd Qu.	Max
Original data	1.0000	3.2500	5.5000	5.5000	7.7500	10.0000
Minmax	0.0000	0.2500	0.5000	0.5000	0.7500	1.0000
Max	0.1000	0.3250	0.5500	0.5500	0.7750	1.0000
Sum	0.0182	0.0591	0.1000	0.1000	0.1409	0.1818
Vector	0.0510	0.1656	0.2803	0.2803	0.3950	0.5096
Logarithmic	0.0000	0.0775	0.1126	0.1000	0.1355	0.1524

to the quotient of the minimum value derived from the original data by the maximum. The other methods narrow this range, as no values appear around 1.

Table 1 presents statistical data of the results. It can be seen that, unlike other methods, in the logarithmic method the median is higher than the mean and is closer to the upper end of the range.

3.2 *Set with Random Natural Numbers*

The second set presents a random natural numbers from the range 50–100. Figure 3 shows the results of calculation. The statistical parameters for the data from this set can be found in Table 2.

The results from the minimum–maximum method are again very similar to the original in terms of shape. The maximum method also preserves the approximate

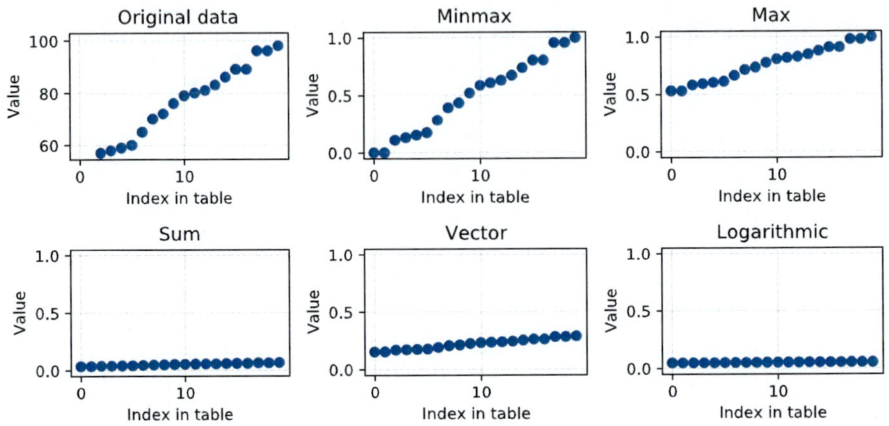

Fig. 3 Visualization of data set Sect. 3.2 before and after normalization

Table 2 Statement of statistical parameters for data set Sect. 3.2

Normalization method	Min	1st Qu.	Median	Mean	3rd Qu.	Max
Original data	52.0000	59.7500	77.5000	74.9000	86.7500	98.0000
Minmax	0.0000	0.1685	0.5543	0.4978	0.7554	1.0000
Max	0.5306	0.6097	0.7908	0.7643	0.8852	1.0000
Sum	0.0347	0.0399	0.0517	0.0500	0.0579	0.0654
Vector	0.1523	0.1750	0.2270	0.2194	0.2541	0.2871
Logarithmic	0.0460	0.0476	0.0506	0.0500	0.0519	0.0534

shape of the data, but compressed in the range from about 0.5–1. The other methods reflect the growing nature of the sorted data but flatten their shape.

As the lower data limit is far from zero, the average data after normalization with the maximum method is closer to 1 than with the minimum–maximum method, as can be seen in the box plot (Fig. 4).

3.3 *Set with Negative Numbers*

The next set on which the normalization methods are tested were negative numbers from −100 to −50. For this reason, it was not possible to apply logarithmic normalization. Figure 5 shows that the minimum–maximum method once again returned data in the range from 0 to 1 with a shape similar to the original.

The results from the maximum method go beyond the scale. Interestingly, in the case of the sum method, the ranking was reversed. The data were flattened, similarly in the vector method. Table 3 and Fig. 6 represent the statistical data received. The

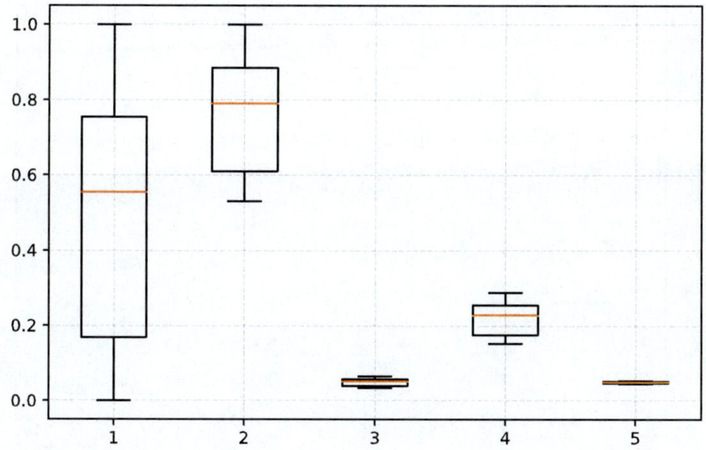

Fig. 4 Box plot for data set Sect. 3.2, where 1—minmax, 2—max, 3—sum, 4—vector, and 5—logarithmic

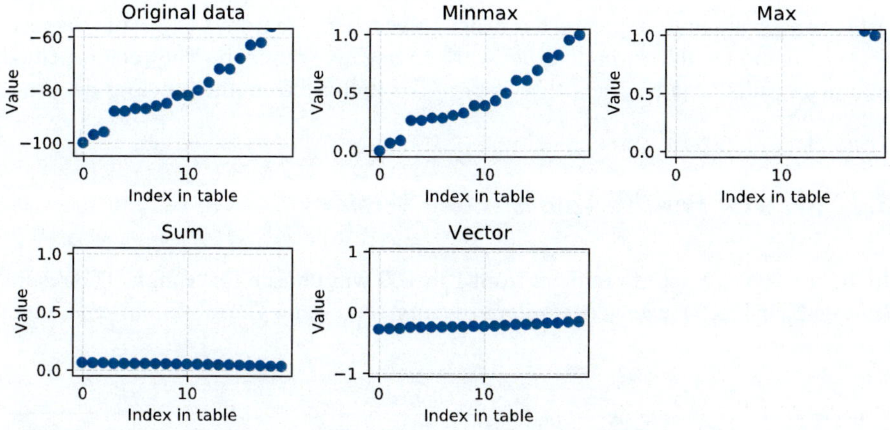

Fig. 5 Visualization of data set Sect. 3.3 before and after normalization

Table 3 Statement of statistical parameters for data set Sect. 3.3

Normalization method	Min	1st Qu.	Median	Mean	3rd Qu.	Max
Original data	−100.0000	−87.2500	−82.0000	−79.1000	−71.0000	−54.0000
Minmax	0.0000	0.2772	0.3913	0.4543	0.6304	1.0000
Max	1.0000	1.3148	1.5185	1.4648	1.6157	1.8519
Sum	0.0341	0.0449	0.0518	0.0500	0.0552	0.0632
Vector	−0.2790	−0.2434	−0.2287	−0.2207	−0.1981	−0.1506

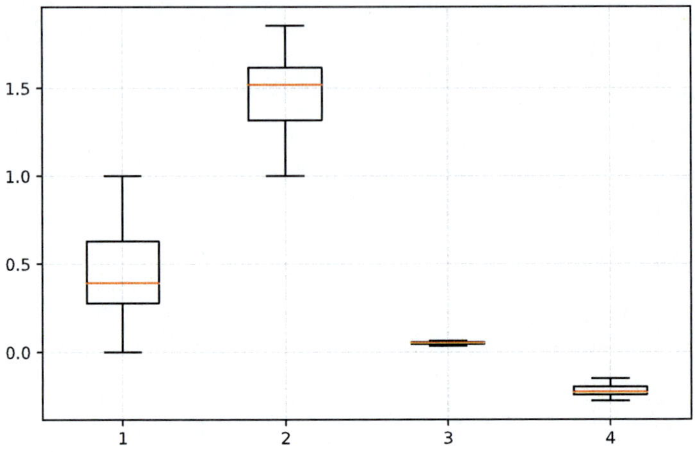

Fig. 6 Box plot for data set Sect. 3.3, where 1—minmax, 2—max, 3—sum, and 4—vector

data from the maximum method has been scaled to a range from 1 to 2. This suggests that the method in this form does not work on negative numbers. The vector method returned results negative, but otherwise similar to those from the previous data set.

3.4 *Set with Positive Values (Long Version)*

In the next case, a set of numbers from 1 to 100 was chosen for testing. The results (presented in Fig. 7) were very similar to the first set of data. The minimum–maximum

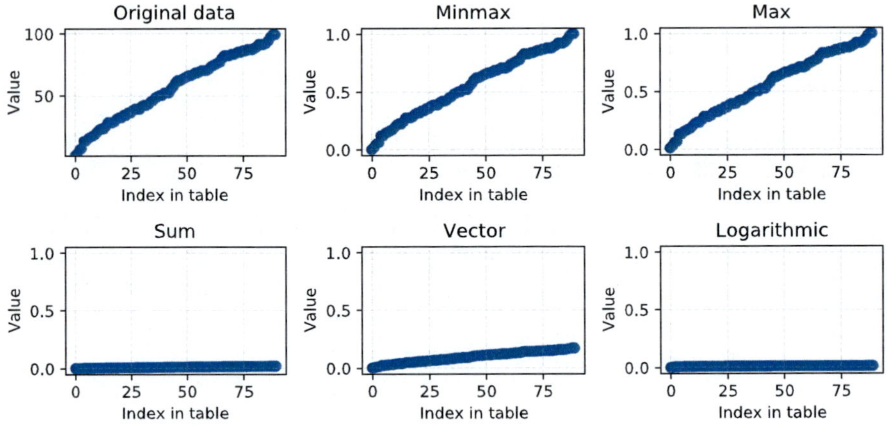

Fig. 7 Visualization of data set Sect. 3.4 before and after normalization

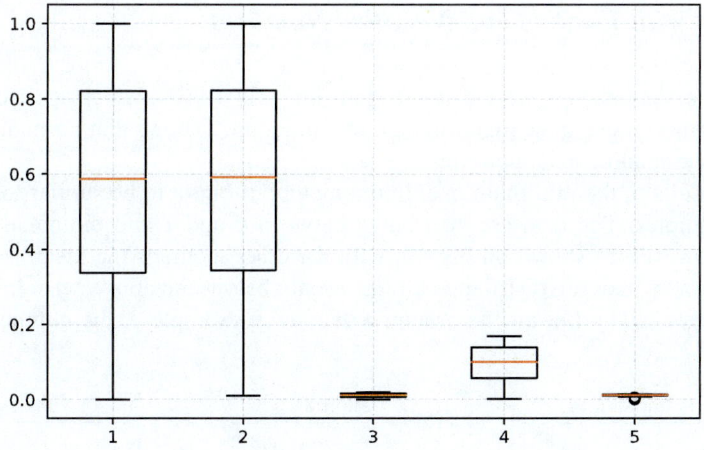

Fig. 8 Box-plot for data set Sect. 3.4, where 1—minmax, 2—max, 3—sum, 4—vector and 5—logarithmic

Table 4 Statement of statistical parameters for data set Sect. 3.4

Normalization method	Min	1st Qu.	Median	Mean	3rd Qu.	Max
Original data	1.0000	34.0000	58.5000	55.5000	81.5000	99.0000
Minmax	0.0000	0.3367	0.5867	0.5561	0.8214	1.0000
Max	0.0101	0.3434	0.5909	0.5606	0.8232	1.0000
Sum	0.0002	0.0068	0.0117	0.0111	0.0163	0.0198
Vector	0.0017	0.0582	0.1001	0.0950	0.1395	0.1695
Logarithmic	0.0000	0.0103	0.0118	0.0111	0.0128	0.0134

method retained the shape of the data; the maximum method retained the shape and had a minimum above zero; the sum, vector, and logarithmic methods flattened the data.

However, there are differences between the first and this set of data, which can be seen in the Fig. 8 and more clearly in the Table 4. Minimum–maximum and maximum methods return more similar averages because the maximum method has the lowest value closer to zero. This is due to the higher upper limit of the original set. Besides, it is worth looking at the results of the three other methods. Both in the first example and here they are flattened, but here they are compressed to an even smaller extent.

3.5 Set with Positive and Negative Numbers

In this case, the numbers were in the range from −100 to 100, which again excluded the possibility of using the logarithmic method. Figure 9 shows that normalized data retains its ascending and approximate shape.

The results of the minimum–maximum method turn out to be similar to the previous examples. The data are distributed between 0 and 1 and the mean is close to 0.5. However, as shown on Fig. 10, with the other methods the mean is close to zero, and the values are distributed almost evenly below and above zero. In the case of the vector method, it can be assumed that, as in example 3, the numbers retain

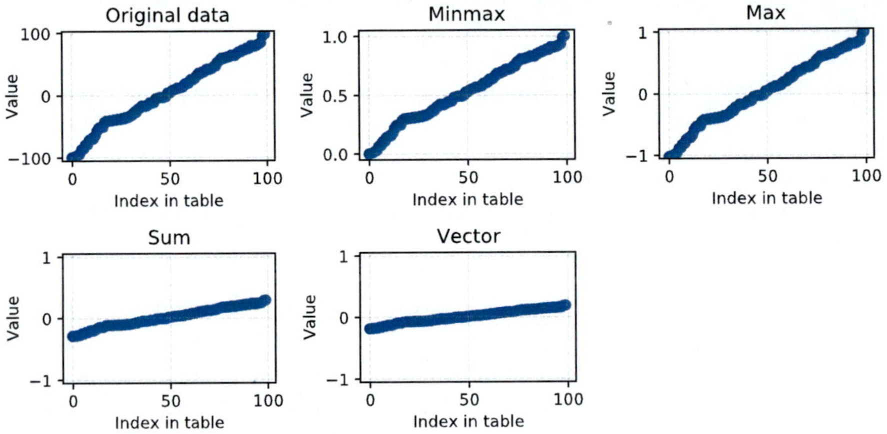

Fig. 9 Visualization of data set Sect. 3.5 before and after normalization

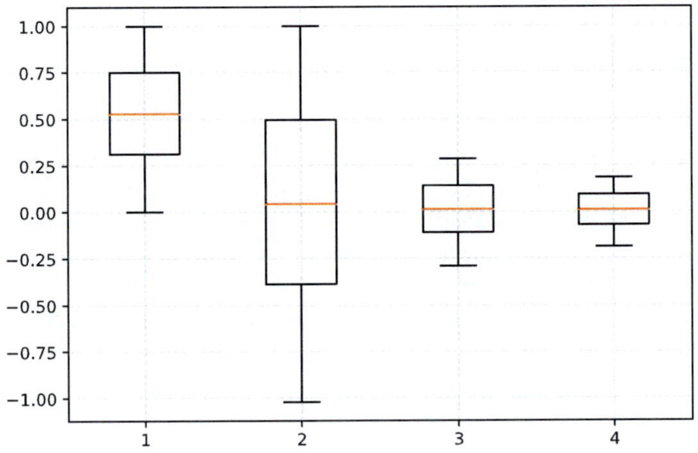

Fig. 10 Box-plot for data set Sect. 3.5, where 1—minmax, 2—max, 3—sum and 4—vector

Table 5 Statement of statistical parameters for data set Sect. 3.5

Normalization method	Min	1st Qu.	Median	Mean	3rd Qu.	Max
Original data	−100.0000	−38.0000	4.0000	3.4400	48.5000	98.0000
Minmax	0.0000	0.3131	0.5253	0.5224	0.7500	1.0000
Max	−1.0204	−0.3878	0.0408	0.0351	0.4949	1.0000
Sum	−0.2907	−0.1105	0.0116	0.0100	0.1410	0.2849
Vector	−0.1882	−0.0715	0.0075	0.0065	0.0913	0.1844

their sign. The statistical parameters are listed in Table 5. Only the results from the minimum–maximum method have all positive values.

3.6 Set with Strongly Asymmetrical Positive Values

In this case, the numbers were in the range from 1 to 10^8, which shows the better possibility of using the logarithmic method. Figure 11 shows that normalized data retains its ascending and approximate shape.

The results of the minimum–maximum and maximum methods turn out to be similar to the shape of original data. The data are distributed between 0 and 1 and the mean is close to 0.2. However, as shown on Fig. 12, with the other methods the mean is close to 0, and the two values outliers are observed. In the case of the vector method, it can be assumed that a less similar to minimum–maximum, and a sum method less than vector method. However, logarithmic method has not outliers and

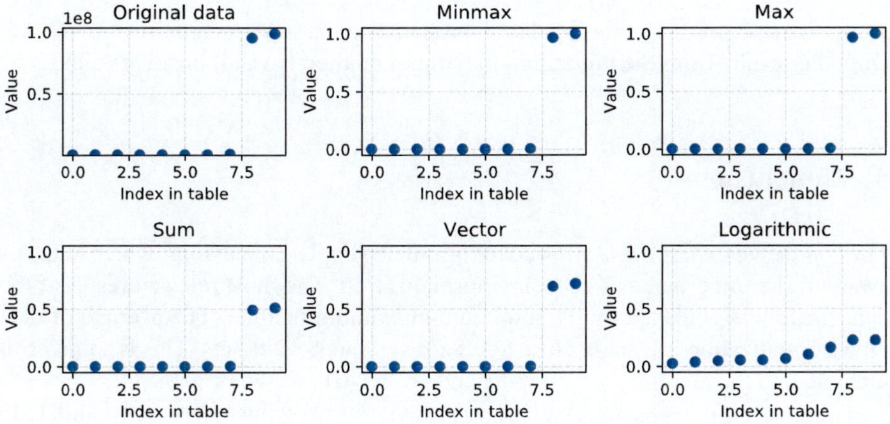

Fig. 11 Visualization of data set Sect. 3.6 before and after normalization

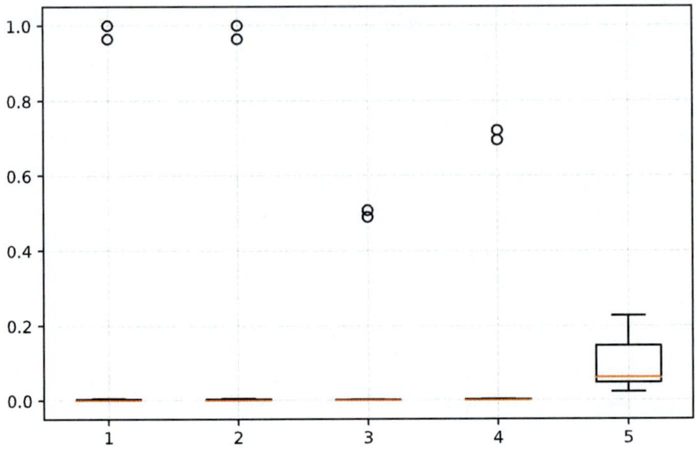

Fig. 12 Box plot for data set Sect. 3.6, where 1—minmax, 2—max, 3—sum, 4—vector, and 5—logarithmic

Table 6 Statement of statistical parameters for data set Sect. 3.6

Normalization method	Min	1st Qu.	Median	Mean	3rd Qu.	Max
Original data	7.2027	54.8640	159.6934	19353933.7089	385055.3206	98257966.8459
Minmax	0.0000	0.0000	0.0000	0.1970	0.0039	1.0000
Max	0.0000	0.0000	0.0000	0.1970	0.0039	1.0000
Sum	0.0000	0.0000	0.0000	0.1000	0.0020	0.5077
Vector	0.0000	0.0000	0.0000	0.1418	0.0028	0.7198
Logarithmic	0.0243	0.0488	0.0612	0.1000	0.1465	0.2262

seems to preferably smooths the data. The statistical parameters are listed in Table 6. Only the results from the minimum–maximum method have all positive values.

4 Conclusions

The conducted experiments allowed to formulate simple conclusions, which make us aware of the importance of choosing normalization. The most relevant result of the data shape was achieved by the minimum–maximum method. The average of data after normalization is usually close to 0.5 for symmetric data sets. This is a universal method, and its operation does not depend on the data we work with.

The minimum–maximum technique always returns values between 0 and 1. In some cases, a zero value can be undesirable. Then the maximum method may be a good choice. Using this approach, the original data is scaled to a range of $[\frac{min(x)}{max(x)}, 1]$.

However, the operation is most stable for positive values. The results will be close to minimum–maximum minimization the more the minimum value is close to zero. The normalization of totals is quite interesting, where the larger the sum of the data set, the smaller the spread after normalization will be. The ranking is either preserved (for positive data) or reversed (for negative data).

Vector normalization is characterized by a lower spread of data after normalization (smoothing). Unfortunately, negative numbers will be negative after normalization. This technique smoothes the data but keeps the trend and would work well for data with noise. The last normalization is logarithmic normalization, which can be used only on positive numbers without zero. This approach smoothes the data and should be used mainly for huge numbers, which are usually presented on a logarithmic scale.

The following study was limited to the most popular normalization methods only. Future directions of the research assume a thorough investigation of the influence of normalization on particular MCDA methods and then extend the research to other possible normalization methods. Also, the interval normalization and fuzzy numbers should be investigated.

References

1. Chakraborty, S., Yeh, C.-H.: A simulation based comparative study of normalization procedures in multiattribute decision making. In: Proceedings of the 6th Conference on 6th WSEAS International Conference on Artificial Intelligence, Knowledge Engineering and Data Bases, vol. 6, pp. 102–109 (2007)
2. Chakraborty, S., Yeh, C.-H.: A simulation comparison of normalization procedures for TOPSIS. In: 2009 International Conference on Computers & Industrial Engineering, pp. 1815–1820. IEEE (2009)
3. Jahan, A., Edwards, K.L.: A state-of-the-art survey on the influence of normalization techniques in ranking: Improving the materials selection process in engineering design. Mater. Des. (1980-2015) **65**, 335–342 (2015)
4. Kizielewicz, B., Wątróbski, J., Sałabun, W.: Identification of relevant criteria set in the MCDA process-wind farm location case study. Energies **13**(24), 6548 (2020)
5. Mathew, M., Sahu, S., Upadhyay, A.K.: Effect of normalization techniques in robot selection using weighted aggregated sum product assessment. Int. J. Innov. Res. Adv. Stud. **4**(2), 59–63 (2017)
6. Milani, A.S., Shanian, A., Madoliat, R., Nemes, J.A.: The effect of normalization norms in multiple attribute decision making models: a case study in gear material selection. Struct. Multidiscip. Optim. **29**(4), 312–318 (2005)
7. Papathanasiou, J., Ploskas, N.: Multiple criteria decision aid (2018)
8. Paradowski, B., Więckowski, J., Dobryakova, L.: Why TOPSIS does not always give correct results? Procedia Comput. Sci. **176**, 3591–3600 (2020)
9. Pavličić, D.: Normalization affects the results of MADM methods. Yugosl. J. Oper. Res. **11**(2), 251–265 (2001)
10. Podvezko, V.: The comparative analysis of MCDA methods SAW and COPRAS. Eng. Econ. **22**(2), 134–146 (2011)
11. Podviezko, A.: Distortions introduced by normalisation of values of criteria in multiple criteria methods of evaluation. LMD DARB **55**, 51–56 (2014)
12. Sałabun, W., et al.: How the normalization of the decision matrix influences the results in the VIKOR method? Procedia Comput. Sci. **176**, 2222–2231 (2020)

13. Sałabun, W., Wątróbski, J., Shekhovtsov, A.: Are MCDA methods benchmarkable? a comparative study of TOPSIS, VIKOR, COPRAS, and PROMETHEE II methods. Symmetry **12**(9), 1549 (2020)
14. Shekhovtsov, A., Sałabun, W.: A comparative case study of the VIKOR and TOPSIS rankings similarity. Procedia Comput. Sci. **176**, 3730–3740 (2020)
15. Vafaei, N., Ribeiro, R.A., Camarinha-Matos, L.M.: Normalization techniques for multi-criteria decision making: analytical hierarchy process case study. In: Doctoral Conference on Computing, Electrical and Industrial Systems, pp. 261–269. Springer (2016)
16. Yazdani, M., Jahan, A., Zavadskas, E.K.: Analysis in material selection: influence of normalization tools on COPRAS-G. Econ. Comput. Econ. Cybern. Stud. Res. **51**(1), (2017)
17. Zardari, N.H., Ahmed, K., Shirazi, S.M., Yusop, Z.B.: Weighting Methods and Their Effects on Multi-criteria Decision Making Model Outcomes in Water Resources Management. Springer, Berlin (2015)

Bipolar Multicriteria Aggregation-Disaggregation Robustness Approach: Theory and Application on European e-Government Benchmarking

Eleftherios Siskos, Giannis Kourousias, and Yannis Siskos

Abstract The aggregation-disaggregation approach is considered as an important tool at the disposal of decision analysts and decision-makers when addressing multiple criteria decision-making problems. This paper proposes a bipolar robustness control approach, implemented in conjunction with the UTASTAR method, where the multicriteria evaluation model is an additive value function. The disaggregation pole of this new algorithm measures and controls the robustness of the evaluation model, as inferred by the decision-maker's preference statements, while the aggregation pole assesses the stability of the results. The bipolar robustness control is complemented with several visualization measures and robustness indicators, the fulfilment of which guarantees the soundness of the model and validates its results. In the end, the methodology is applied to the problem of e-government readiness evaluation in Europe, resulting in the ranking of 22 European countries.

Keywords Multiple criteria · Aggregation-disaggregation approach · e-government · Robustness analysis

Abbreviations

ARP	Average Range of Preferential Parameters
ARRI	Average Range of the Ranking Index
ASI	Average Stability Index
AVG	Average

E. Siskos · G. Kourousias
School of Electrical and Computer Engineering, National Technical University of Athens, 9, Iroon Polytechniou Str., 15780 Zografou, Athens, Greece
e-mail: lsiskos@epu.ntua.gr

G. Kourousias
e-mail: kourousias@protonmail.com

Y. Siskos (✉)
Department of Informatics, University of Piraeus, 80 Karaoli & Dimitriou str., 18534 Piraeus, Greece

A. J. Kulkarni (ed.), *Multiple Criteria Decision Making*, Studies in Systems, Decision and Control 407, https://doi.org/10.1007/978-981-16-7414-3_7

DM	Decision-Maker
ERA	Extreme Ranking Analysis
GDP	Gross Domestic Product
ICT	Information and Communications Technology
LP	Linear Programming
R&D	Research and Development
RARR	Ratio of the Average Range of the Ranking
ROR	Robust Ordinal Regression
SPRI	Statistical Preference Relations Index
UTA	UTilités Additives

1 Introduction

Aggregation-disaggregation or ordinal regression approach is currently considered as an important tool at the disposal of potential analysts and decision-makers when addressing decision-making problems under the regime of multiple criteria. The main goal of this philosophy lies in the assessment/inference of preference models from given preferential structures and the support of decision-aiding activities through operational models, within the aforementioned framework (see [9, 12, 14]).

The most representative disaggregation method is the UTA method (UTilités Additives), proposed by [8]. It aims at inferring one or more additive value functions from a given ranking or other preference statements (e.g., pairwise comparisons) made on a reference set of alternatives A_R. The method uses special LP (Linear Programming) techniques to assess these functions so that the ranking(s) obtained through these functions on A_R is (are) as consistent as possible with the reference preference statements. An improved version of the original UTA method is UTASTAR method, which in presented in Sect. 2 and more explicitly in the Appendix B of this paper.

Recently, special attention has been given to certain robustness issues of the ordinal regression framework and the way that the preferential parameters are estimated through the UTA-type inference engine [5, 11, 12]. Towards this direction, this paper proposes an interactive bipolar robustness control procedure to strengthen the implementation of the UTASTAR method and the additive value function, as part of the former. The disaggregation pole of this algorithm measures and controls the robustness of the evaluation model, as inferred by the decision-maker's (DM's) preference statements, while the aggregation pole assesses and validates the ranking results, given by the model. The methodology is complemented with certain visualization measures and robustness indicators, which drive and guide the procedure, until the acquisition of secure and robust results.

In the end, the proposed robustness control methodology is implemented in the context of a real-world case study, in order to stress test and verify its validity and efficacy. Specifically, the UTASTAR method, coupled with the robustness control methodology, is applied to the evaluation of the e-government performance in Europe. The evaluation model built with the aid of the UTASTAR method is applied for the ranking of 22 European countries. This application is based on the multicriteria e-government modeling work of [13] and considers updated data on the countries.

The paper is organized as follows: Sect. 2 presents the theoretical basis of the additive value model and the UTASTAR framework. The proposed bipolar robustness control procedure is outlined in Sect. 3. Section 4 briefly presents the decision model, based on which e-government is assessed at the national level. Sections 5 and 6 are purely practical and implement the robustness control methodology to rank 22 European countries over their e-government performance. Finally, Sect. 7 concludes the paper.

2 Additive Value Model and UTASTAR Method

2.1 Problem Statement and Notation

The most common approach for evaluating a set of actions $A = \{a, b, c, \ldots\}$ is to use an additive representation. The multicriteria evaluation model proposed in this paper is an additive value function u, described by the following formulae:

$$u(g) = \sum_{i=1}^{n} p_i u_i(g_i) \tag{1}$$

$$u_i(g_{i*}) = 0,\ u_i\left(g_i^*\right) = 1,\ \text{for}\ i = 1, 2, \ldots, n \tag{2}$$

$$\sum_{i=1}^{n} p_i = 1 \tag{3}$$

$$p_i \geq 0,\ \text{for}\ i = 1, 2, \ldots, n \tag{4}$$

where $\mathbf{g} = (g_1, g_2, \ldots, g_n)$ is the performance vector of an action on n criteria; g_{i*} and g_i^* are the least and most preferable levels of the criterion g_i, respectively; $u_i(g_i)$, $i = 1, 2, \ldots, n$ are non-decreasing marginal value functions of the performances g_i, $i = 1, 2, \ldots, n$; and p_i is the relative weight of the *i-th* function $u_i(g_i)$.

Thus, for a given action a, $\mathbf{g}(a)$ and $u\left[\mathbf{g}(a)\right]$ represent the multicriteria vector of performances and the global value (in the interval [0, 1]) of the action a, respectively.

Both the marginal and the global value functions have the monotonicity property of the true criterion. For two actions a and b, the following properties hold:

$$u\left[\mathbf{g}(a)\right] > u\left[\mathbf{g}(b)\right] \Leftrightarrow a > b\ \text{(Preference)} \tag{5}$$

$$u\big[\mathbf{g}(a)\big] = u\big[\mathbf{g}(b)\big] \Leftrightarrow a \sim b \text{ (indifference)} \tag{6}$$

The necessary hypothesis to validate an additive value function for a given DM is the preference independence of the criteria (see [3], for instance).

2.2 The UTASTAR Ordinal Regression Method

Jacquet-Lagrèze and Siskos [9] addressed the condition of preference independence by suggesting an ordinal regression or disaggregation approach, which aims at inferring one or more additive value functions from given DM's preference statements (see also [14]). More specifically, the UTASTAR algorithm (see Appendix B) infers additive value functions from a ranking expressed on a reference set of actions A_R. The method uses special LP techniques to assess these functions so that the ranking(s) obtained through these functions on A_R is (are) as consistent as possible with the DM's preference ranking.

In UTA methods, the additive value model to be assessed has the following unweighted form, which is strictly equivalent to the above weighted form (1)–(4):

$$u(\mathbf{g}) = \sum_{i=1}^{n} u_i(g_i) \tag{7}$$

subject to normalization constraints:

$$\begin{cases} \sum_{i=1}^{n} u_i(g_i^*) = 1 \\ u_i(g_{i*}) = 0 \;\; \forall i = 1, 2, \ldots, n \end{cases} \tag{8}$$

where $u_i, i = 1, 2, \ldots, n$ are non-decreasing real valued functions, named marginal value functions.

In UTA methods (see [14]), each value functions u_i is supposed to have a piecewise linear form on a_i points of the corresponding evaluation scale $[g_{i*}, g_i^*]$. This presupposes that each scale is already discretized into $a_i - 1$ equally distant intervals.

In addition, the monotonicity constraints are taken into account, with the aid of the following transformations:

$$w_{ij} = u_i\left(g_i^{j+1}\right) - u_i\left(g_i^{j}\right) \geq 0 \, \forall i = 1, 2, \ldots, n \text{ and } j = 1, 2, \ldots, a_i - 1 \tag{9}$$

Thus, the monotonicity conditions may be replaced by the non-negative constraints for the variables w_{ij} (a_i is the number of points, on which the value function u_i is assessed). Finally, the determination of the value function u_i is fully achieved, when all the values of the w_{ij} variables, whose number is $\sum_{i=1}^{n}(\alpha_i - 1)$, have been determined.

3 The Bipolar Ordinal Regression Process of Robustness Control

3.1 Principles

The effectiveness of the UTASTAR inference engine depends on the size of the DM's preference information, included in the reference set, that is, more specifically, on the number of reference actions that constitute the set A_R. This means that the larger the reference set, the more "accurate" the additive value model, estimated by the method.

UTASTAR algorithm (see also Appendix B) shows that the DM's evaluation model may not be a unique additive value function but a set of functions, all being compatible with the preference statements provided to the analyst. This infinite set of functions comprises a polyhedral set, confined under some linear constraints, in the $\sum_{i=1}^{n}(\alpha_i - 1)$ dimension space. Greco et al. [5] proposed a general methodological framework, named "Robust Ordinal Regression" (ROR), which can be implemented synergistically to the disaggregation methods and aims at enhancing the robustness of the estimated results. ROR is based on the principle, according to which the decisions and proposals emerge after considering all those parameters that are compatible with the preferences of the DM.

Towards this direction, a robustness control algorithm is proposed, in order to examine, analyze, measure, and assess the robustness of the decision-making procedure. This algorithm focuses separately in the two different aspects/poles of the procedure, namely, the disaggregation and the aggregation one (bipolar procedure).

Figure 1 outlines graphically the flowchart of the interactive bipolar robustness control, which manages robustness in both phases/poles of the decision support process. The robustness control process is initiated with the inference of the additive value model, resulting from the ranking of the reference actions. It then proceeds to the assessment of the robustness of the model, with the option of discontinuing the modeling process, if the results are not satisfactory. In this case, the analyst asks the DM to enrich the reference set with additional reference actions or add other new preference statements.

Alternatively, the process moves from the disaggregation to the aggregation pole, where the additive value model is implemented and the ranking of the real actions is achieved. Robustness is again measured in this pole, in terms of the stability of the ranking positions of each action. If the robustness of the results is adequate to support a safe decision, the algorithm is terminated, otherwise the analyst returns to

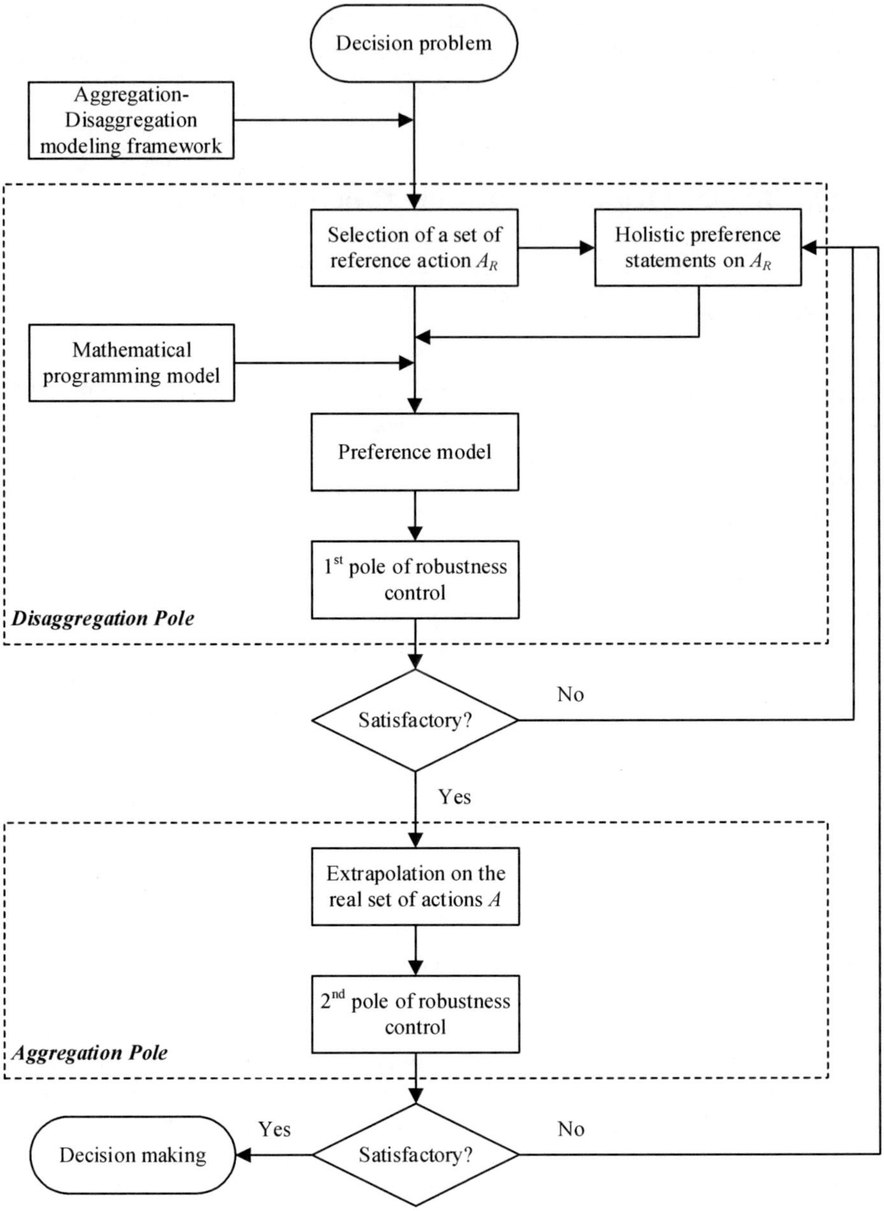

Fig. 1 The bipolar robustness control procedure

the disaggregation pole and asks the DM for the acquisition of additional preferential information.

The inclusion of the DM in the whole procedure is significant, since his/her preference input is constantly required, in order to improve the stability of either the decision model or the evaluation results. On the other hand, the DM is, during the complete implementation and the rounds of feedback, fully aware of the modeling stages, the impact of his/her input on the results, the evolution of the robustness indicators, and is deciding, in cooperation with the analyst, whether the robustness in either of the poles is satisfactory.

3.2 Robustness Control Measures

Various numerical indicators and visualization tools are used to assess the robustness of the parameters calculated through the UTASTAR method. These indices are included and applied in either the disaggregation or the aggregation pole of Fig. 1. Some of these indices are reported here, especially those used in the problem of e-government evaluation in the next three sections.

Disaggregation pole

Two main indices can be recognized in this category. The use of these indices presupposes the production of multiple sets of preferential parameters. A usual way to achieve this, when implementing the UTASTAR method is the max–min LPs technique (see step 4 in Appendix B). During this procedure, all or a subset of model's parameters are successively minimized and maximized, under the set of feasibility constraints, and then visualized.

Let p_{rs} be the set of the model's preference parameters, produced by the UTASTAR algorithm, where r denotes a specific instance, in which the parameter p is estimated ($r = 1, 2, \ldots, R$) and s denotes a specific parameter ($s = 1, 2, \ldots, S$). In UTASTAR method, where w_{ij} are to be estimated during step 4, the number of instances is $R = 2\sum_{i=1}^{n}(\alpha_i - 1)$ and the number of different parameters w_{ij} is $S = \sum_{i=1}^{n}(\alpha_i - 1)$.

Average Range of Preferential Parameters (ARP)

The calculation of the *ARP* requires the priori implementation of the max–min LPs technique and is defined as follows:

$$ARP = \frac{1}{S}\sum_{S=1}^{S}\left[\max_r(p_{rs}) - \min_r(p_{rs})\right] \tag{10}$$

where p_{rs} is the r-th instance of the s-th preferential parameter. This index ranges in [0, 1] and indicates the average possible variation of the preferential parameters. Therefore, it receives lower values as the robustness of a model increases. *ARP* receives the value of 0 when a unique preference model reflects the preference statements of the DM.

Average Stability Index (ASI)

The average stability index is a robustness index proposed by Siskos and Grigoroudis [6] and indicates the average value of the normalized standard deviation of the preferential parameters. *ASI* also ranges in [0, 1] and returns the value of 1, when perfect robustness is achieved.

In the presented approach, *ASI* has the following form:

$$ASI = 1 - \frac{1}{\sum_{i=1}^{n}(\alpha_i - 1)} \sum_{i=1}^{n}\sum_{j=2}^{\alpha_i} \frac{\sqrt{\sum_{i=1}^{n}(\alpha_i - 1)\sum_{r=1}^{R}(u_{ij}^{r})^2 - \left(\sum_{r=1}^{R} u_{ij}^{r}\right)^2}}{2\sqrt{\sum_{i=1}^{n}\alpha_i - (n+1)}} \tag{11}$$

where u_{ij}^{r} is the r-th instance of u_{ij} during the max–min LPs procedure with $R = 2\sum_{i=1}^{n}(\alpha_i - 1)$.

Most representative preference model

Finally, an average additive value model ("barycenter") of all different instances should be obtained, as the most representative preference model solution in the corresponding hyper-polyhedron of all possible solutions (see also [11] about the specification of representative parameter sets). This average additive model accommodates the average values of all the preference parameters (resulting for the max–min LPs procedure here) and is used for the acquisition of a representative ranking of the actions under evaluation.

Aggregation pole

The exploitation of the indices related to the disaggregation pole offers a comprehensive view of the robustness of the decision model, though it does not guarantee robust results after the its implementation. The use of appropriate indices in the aggregation pole (2nd pole) is therefore necessary to monitor robustness as the algorithm progresses. Again, these indices work under the condition that certain techniques are implemented.

Extreme ranking analysis

The extreme ranking analysis, proposed by Kadzinski et al. [10], uses mixed-integer linear programming techniques, in order to estimate every action's best and worst possible position in the ranking. It constitutes an additional tool to visualize the variation of actions' ranking.

Average Range of the Ranking (ARRI) and Ratio of the Average Range of the Ranking (RARR)

The average range of the ranking index and the ratio of the average range of the ranking are two indices, used in conjunction with the Extreme Ranking Analysis. Specifically, *ARRI* depicts the possible number of positions that an average action can occupy in the whole ranking, while *RARR* reflects the ratio of the aforementioned deviation, with respect to the whole number of the alternatives under evaluation. The optimal values of *ARRI* and *RARR* are 1 and 0%, respectively, and they are calculated using the following formulae:

$$ARRI = \frac{1}{m}\sum_{k=1}^{m}\left(\left|R_*(k) - R^*(k)\right| + 1\right) \tag{12}$$

$$RARR = \frac{ARRI - 1}{m - 1} \cdot 100\% \tag{13}$$

where $R_*(k)$ and $R^*(k)$ are the worst and best possible ranking positions, respectively, for the k-th alternative, while m is the number of all the alternatives under evaluation.

Statistical Preference Relations Index (SPRI)

The *SPRI* offers a comprehensive way to examine the stability of the ranking positions, achieved by the whole entity of actions. The calculation of *SPRI* prerequisites the implementation of a random sampling technique, such as the Hit-and-Run algorithm [15] and generally methods that generate a statistically adequate number of model parameters, within the polyhedron of parameters. Then, for each vector of parameters, the associated ranking is calculated, using the outranking method.

Building on these multiple rankings, *SPRI* calculates the separate frequencies or probabilities that each action occupies a single ranking position in the final ranking, and constitutes a measure, providing clear insight on the robustness of the results. Specifically, the probability that an action a_k is ranked in the t-th position is calculated using the following relation:

$$P_t^k = \frac{c_t^k}{m} \cdot 100\% \tag{14}$$

where c_t^k is the number of samples/instances that position an action a_k in the tth position, $t = 1, 2, \ldots, N$ and m is the number of all the samples/instances.

The statistical preference relations index is then calculated using the following equation:

$$SPRI = \frac{1}{m}\sum_{k=1}^{m}\sum_{t=1}^{N} P_t^k \tag{15}$$

SPRI reaches the optimal value of 100% when each action occupies a single ranking position, with a statistical probability of 100%. In other words, the same exact ranking occurs, after the implementation of the additive value model, for every different instance/sampling of the preferential parameters.

4 Case Study: Evaluating e-Government Performance in Europe

E-government refers to the introduction of telecommunications and computer technologies in public administration and the new administrative practices that these technologies enable. The purpose and objective of this digitalization is to improve the services provided to citizens and to facilitate the procedures at the intra-administrative level. It promotes, therefore, both the utilization of existing electronic infrastructure and the development of new ones to support the interaction of citizens, businesses, and governmental agencies with the authorities.

At the level of providing services to the citizens, the provision of innovative services is sought, through a single point of provision (portal). Its goal is to gather all the individual services in one place, where access can be achieved by multiple means (computer, mobile devices, etc.). In this way, the effective servicing of citizens is achieved with great transparency and efficiency. For businesses, e-government has a significant impact on day-to-day operations, by allowing the bypassing of dysfunctional bureaucratic procedures, and enabling cost and time efficiencies. From the government's point of view, the benefits presented in the field of public administration are strategic (limiting bureaucracy, improving relations with citizens, etc.), administrative (cost/time reduction, support for partnerships, etc.), and functional (automation of processes, efficient utilization of knowledge/workforce, etc.).

Several scholars and institutions have researched on the benefits of e-government readiness and the tools and methods to evaluate and monitor e-government progress. For detailed views, one could refer to [1, 2, 7], European [4], and [16], for instance. Delving further deep, the evaluation of e-government is essential, in order to measure performance at a global, national, regional, and local level and to highlight possible areas for improvement. Although many relevant studies have been conducted in the past by various organizations (see European [4] and [16], for instance), the multidimensional/multicriteria nature of such assessments has not been fully established yet.

The evaluation of e-government, as part of the application area of this study, is based on the multicriteria evaluation approach proposed by Siskos et al. [13], but emphasizes on the robustness of the modeling work and the results. Siskos et al. [13], in their paper, proposed a multiple criteria evaluation system for the assessment of a country's performance on e-government. The evaluation criteria that were used, are grouped into four points of view:

Table 1 E-government evaluation criteria, indices, criteria ranges, and data sources

Criterion	Metric	Worst level	Best level	Data source
g_1- Access to the web	% population	0	100	Eurostat
g_2- Broadband internet connection	% population	0	100	Eurostat
g_3- Gross domestic product (GDP) on information & communications technology (ICT) and research & development (R&D)	% GDP	0	4	Eurostat
g_4- Online Sophistication	% index	0	100	European Commission
g_5- E-participation	Index [0–1]	0	1	United Nations
g_6- Citizens' online interaction with authorities	% citizens	0	100	Eurostat
g_7- Businesses' online interaction with authorities	% businesses	0	100	Eurostat
g_8- User's experience	% index	0	100	European Commission

i. country infrastructures (two criteria: access to the web; broadband internet connection),
ii. national investments (one criterion: % GDP on information and communications technology and research and development),
iii. e-processes (two criteria: online sophistication; e-participation), and
iv. users' attitude against the e-processes (three criteria: citizens' online interaction with authorities; businesses' online interaction with authorities; user's experience).

A synopsis of the eight evaluation criteria, along with their measurement indices, scoring ranges, and data sources is presented in Table 1. Further details on this evaluation basis can be accessed at [13].

The approach proposed in this paper, evaluates 22 European countries, based on their performance on the aforementioned eight criteria, corresponding to the year of 2017. The country data on the eight criteria are presented in Appendix A.

5 Implementation—Part 1: Initialization and First Robustness Control

5.1 Initialization Phase

The additive value model and the UTASTAR method are now implemented in conjunction with the robustness methodological framework, described in Sect. 3. This Section provides the first attempt to model the e-government additive value system, together with the robustness assessment on the modeling parameters. The decision-maker is an expert of the Laboratory of Decision Systems, of the National Technical University of Athens, with high knowledge in e-government and relevant experience of more than 20 years.

The initialization phase begins with the specification of the reference countries, as mandated by the UTASTAR method, and their ranking by the DM, based on their e-government performance. The reference countries, in this case, are fictitious countries, representative to the European e-government status quo and created by the analyst. Therefore, they do not constitute a selection of real countries among the 22 under evaluation.

After, the DM provides the ranking of the reference countries, the mathematical model of UTASTAR is applied, in order to construct the e-government evaluation model. Subsequently, the model is stress tested by the bipolar robustness control procedure, with a view to assessing its reliability and stability, prior to implementing it on the real country set. It is made clear here, that as bipolar robustness control progresses, and additional preference information is provided by the DM, the transition from the disaggregation to the aggregation pole (and vice versa) leads to better results.

5.2 *Creation and Ranking of the Reference Countries—Phase A*

Ten virtual countries, which are assigned score in each criterion, on the basis of the European countries performances, are used as the reference set for the application of UTASTAR. The way in which the criteria values are assigned to the reference countries is such that, on one hand, they present realistic references to each criterion and, on the other hand, are representative of the European data, including at the same time some few extreme values. Moreover, in order to facilitate the DM in his evaluation, the reference countries are given rounded values to the criteria, as well as many common in between them. The ten reference countries (C1–C10), as well as their ranking by the DM in descending e-government performance, are presented in Table 2.

5.3 *Application of UTASTAR*

In order to apply the UTASTAR method to the data of Table 2, the analyst needs to set the values of the eight parameters a_i, $i = 1, 2, \ldots, 8$, i.e., the points at which the piecewise linear marginal value functions will be delimited, as well as the value δ that expresses the minimum level of preference between two consecutive classes of the ranking. These parameters were set as follows:

Table 2 Ranking of the ten reference countries

Ranking	Reference countries	g_1	g_2	g_3	g_4	g_5	g_6	g_7	g_8
1	C6	85	75	2.5	85	0.8	40	80	60
2	C8	85	75	3.0	65	0.6	50	80	60
3	C7	85	75	2.5	65	0.6	50	85	60
4	C5	95	70	2.0	85	0.3	40	80	60
5	C1	85	70	2.0	75	0.6	40	80	60
6	C3	85	60	2.5	75	0.6	40	80	60
7	C4	95	70	2.0	55	0.4	40	80	60
8	C2	95	60	2.0	75	0.6	40	80	60
9	C9	85	75	3.0	65	0.6	30	70	50
10	C10	80	65	3.0	65	0.6	40	80	40

$a_i = 5$, for every $i = 1, 2, \ldots, 8$ and $\delta = 0.01$

Consequently, the number of variables w_{ij} of the additive value model under development is $8 \times (5 - 1) = 32$, according to the formulas (7) and (8).

To solve the formed linear programs, we used the process outlined below with the help of the mathematical programing platform GAMS IDE. First, we set as objective function the sum of all the errors contained in the inequalities. By minimizing this function, we ensure the absence of logical errors in the ranking given by the DM. If the solution does not result in a zero error, then the reference countries have been ranked in a "non-rational" way, which means that there is no additive value model that can give back this exact ranking. In that case, the analyst intervenes, asking the DM to rearrange the countries that exhibited errors according to the linear problem solution, in such a way that the corresponding errors are zeroed. Consequently, the logical accuracy of the data given as input to the model is ensured, without compromising the preferential data of the DM.

After ensuring that all possible errors are zeroed, we can proceed to determine the values of the parameters w_{ij}, and apply them to the additive value model. To do this, the w_{ij} parameters are consecutively maximized and minimized in the same mathematical programming model, without the errors (max – min process), with a view to subsequently estimating their average levels.

For example, to calculate the range of parameters, associated with the first criterion, we maximize and minimize the following four objective functions:

$$z_{11} = w_{11}$$

$$z_{12} = w_{11} + w_{12}$$

$$z_{13} = w_{11} + w_{12} + w_{13}$$

$$W_1 = z_{14} = w_{11} + w_{12} + w_{13} + w_{14}$$

where W_1 corresponds to the weight of the first criterion. By maximizing and minimizing this function, we therefore estimate the deviation range of the weight of the first criterion.

In this way, a total of 64 runs are performed at this stage (instances—max and min 32 objective functions) to determine the ranges of the weights of the eight criteria. The results obtained after calculating $u_i(g_i)$, through the w_{ij} variables and according to the formula (9) (value variation), are shown in Fig. 2. The dashed line represents the average marginal value functions, as calculated after the end of all the runs. The variation of criteria weights W_i, $i = 1, 2, \ldots, 8$ is also reported in Fig. 3.

5.4 Evaluation of Robustness

Based on Figs. 2 and 3, we easily realize that the range of both the weights W_i and the marginal variables w_{ij} is practically uncontrollable, as they extend to approximately 80% of their feasible space [0–1]. In addition, the average values of all variables generally tend to their minimum values, which does not help differentiate adequately the importance of difference between the criteria.

As a result, the model exhibits insufficient robustness at this stage, which is also confirmed by the relatively low price of 0.885 of the optimistic indicator *ASI*. Under these circumstances, the analyst cannot calculate a representative model and, therefore decides to repeat the process from the beginning of the disaggregation pole, and thus ask the DM with additional information that will help determine more accurately the preference model.

6 Implementation—Part 2: Robust Evaluation of European e-Government

6.1 Phase B—20 Reference Countries

Integration of new countries in the reference ranking

Based on the procedure followed in Phase A, ten additional reference countries (C11–C20) are created by the analyst, according to the rules, requirements, and specifications, described in Sect. 5. The DM is then asked to integrate them into the predetermined ranking of Phase A (see Table 3). The ten new reference countries are shown in bold. It is important to highlight that the relevant positions of the first ten

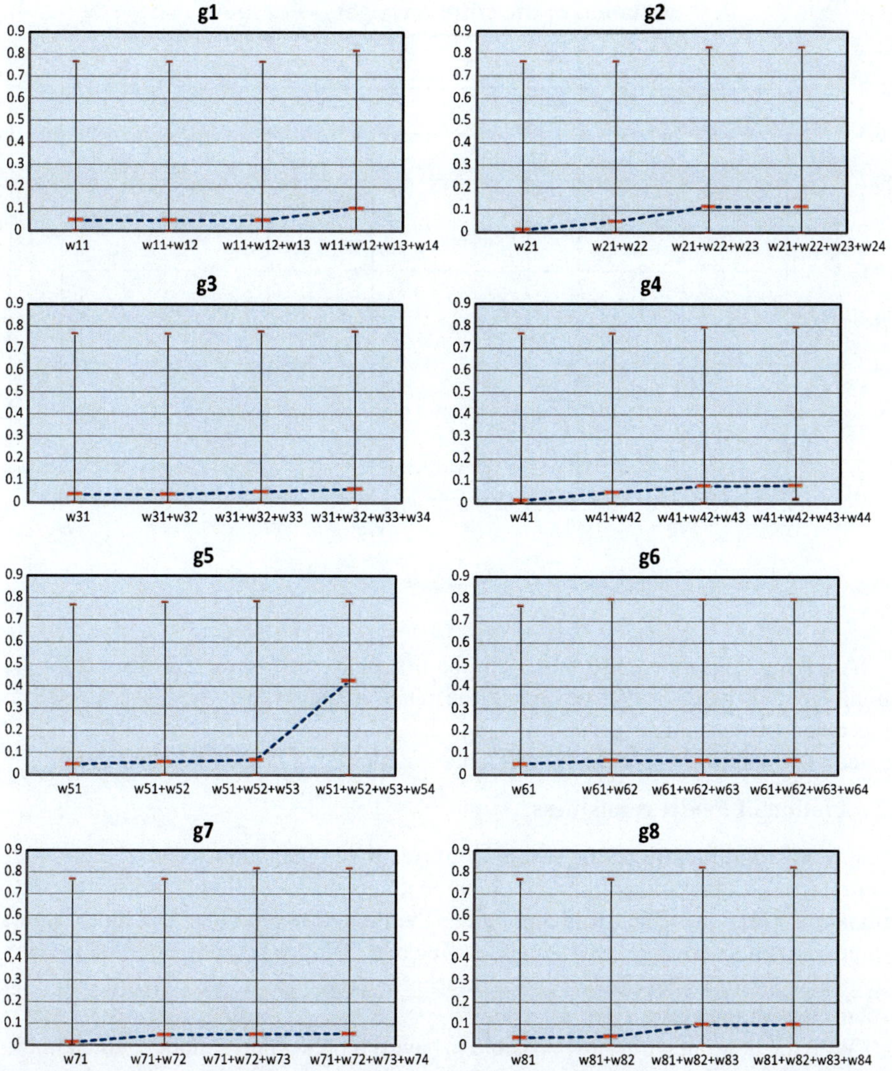

Fig. 2 Marginal value functions of the 10-reference country model

countries, as they were formed in Phase A, cannot be changed, as this would signify a logical discontinuity of the model.

UTASTAR application

Using the new ranking data, 19 inequalities were extracted to form the linear program of UTASTAR. The term δ has now been reduced to 0.005 to ensure the delivery of a reasonable ranking of the countries by the DM.

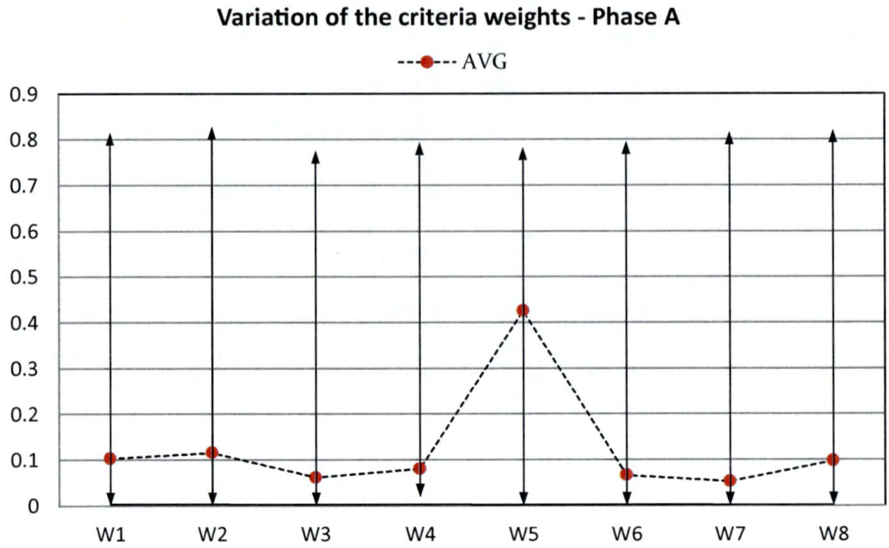

Fig. 3 Variation range of the criteria weights—Phase A

Applying the same constraints as in the first phase and zeroing all the errors, we implemented the 64 runs to estimate the range of the criteria weights, as well as the range of the marginal variables w_{ij}. The range of the criteria weights W_i, as calculated with the aid of the 20-reference country data, is shown below in Fig. 4.

Evaluation of model robustness

The more solid picture of the results confirms the higher robustness of the model, compared to the previous Phase and demonstrates the key role played by the additional information, which was inputted to the model. Indeed, the *ASI* index shows improvement in its value, receiving the value of 0.936. This increase of 0.05 in absolute value, compared to Phase A, can be seen as a major step towards improving the robustness of the system's results.

With the new data, it is now possible to have a representative model, which marks the transition to the aggregation pole. Thus, additional tools will be used, which will assess whether the robustness of the model and the results are satisfactory.

Ranking of European countries and robustness control

By applying the representative model (barycentric/average weighting profile) of this phase to the real country data (see Appendix A), we can use the tools of the aggregation pole to decide on the level of robustness of the results.

Such a tool is the Extreme Ranking Analysis (ERA), which calculates the maximum and minimum possible position that each country can get in the whole ranking. The calculation of the extreme rankings was achieved after solving the relevant problems of mixed-integer programming in the modeling platform GAMS IDE.

Table 3 Integrated ranking of the 20 reference countries

Ranking positions	Fictitious countries	g_1	g_2	g_3	g_4	g_5	g_6	g_7	g_8
1	C6	85	75	2.5	85	0.8	40	80	60
2	C8	85	75	3.0	65	0.6	50	80	60
3	C7	85	75	2.5	65	0.6	50	85	60
4	**C20**	**95**	**75**	**1.0**	**95**	**1.0**	**20**	**60**	**75**
5	**C17**	**85**	**65**	**2.0**	**95**	**0.8**	**40**	**80**	**60**
6	C5	95	70	2.0	85	0.3	40	80	60
7	**C12**	**90**	**70**	**1.0**	**70**	**0.3**	**40**	**85**	**75**
8	**C19**	**80**	**60**	**3.0**	**75**	**0.8**	**60**	**80**	**70**
9	**C14**	**80**	**60**	**2.0**	**80**	**0.8**	**65**	**80**	**60**
10	C1	85	70	2.0	75	0.6	40	80	60
11	**C13**	**85**	**60**	**2.0**	**75**	**0.8**	**50**	**80**	**60**
12	C3	85	60	2.5	75	0.6	40	80	60
13	C4	95	70	2.0	55	0.4	40	80	60
14	C2	95	60	2.0	75	0.6	40	80	60
15	C9	85	75	3.0	65	0.6	30	70	50
16	**C18**	**90**	**70**	**1.0**	**75**	**0.6**	**40**	**70**	**50**
17	C10	80	65	3.0	65	0.6	40	80	40
18	**C16**	**85**	**65**	**2.5**	**80**	**0.4**	**40**	**60**	**40**
19	**C15**	**90**	**70**	**2.0**	**50**	**0.4**	**40**	**60**	**30**
20	**C11**	**80**	**60**	**1.0**	**65**	**0.3**	**20**	**70**	**40**

The results are presented with the help of the diagram in Fig. 5, which shows the position of each country in the representative ranking with a red dot. The vertical arrows depict the range of all the possible positions for a given country.

The visualized results of Fig. 5 reveal a wide range of possible rankings for each country, which is due to the inherent instability of the values of the variables w_{ij}. The *ARRI* robustness index in this case is equal to 6.27, which indicates that the average country has more than six possible ranking positions. Respectively, the *RARR* index gets a value of 24%, which cannot be considered acceptable.

Given the above, the robustness of the model has still not reached an adequate level and, therefore, it is decided to return to the disaggregation pole and repeat the process, as indicated by the bipolar robustness control methodology.

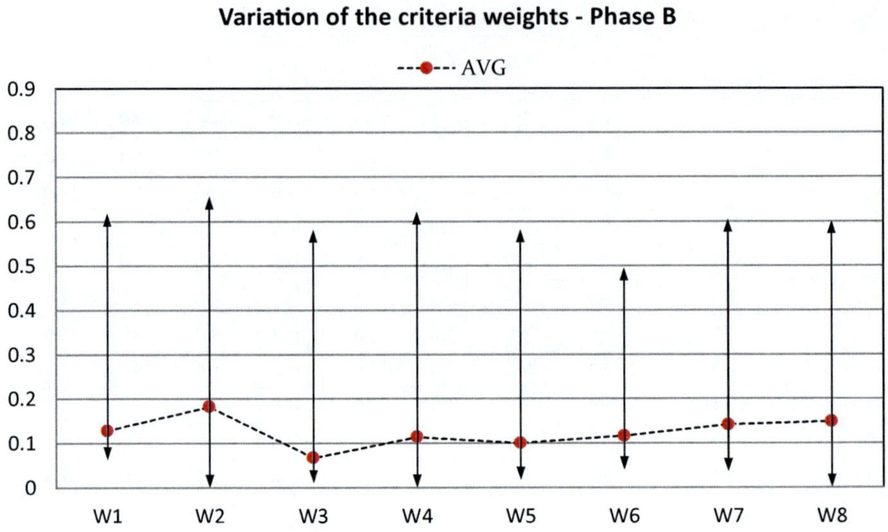

Fig. 4 Variation range of the criteria weights—Phase B

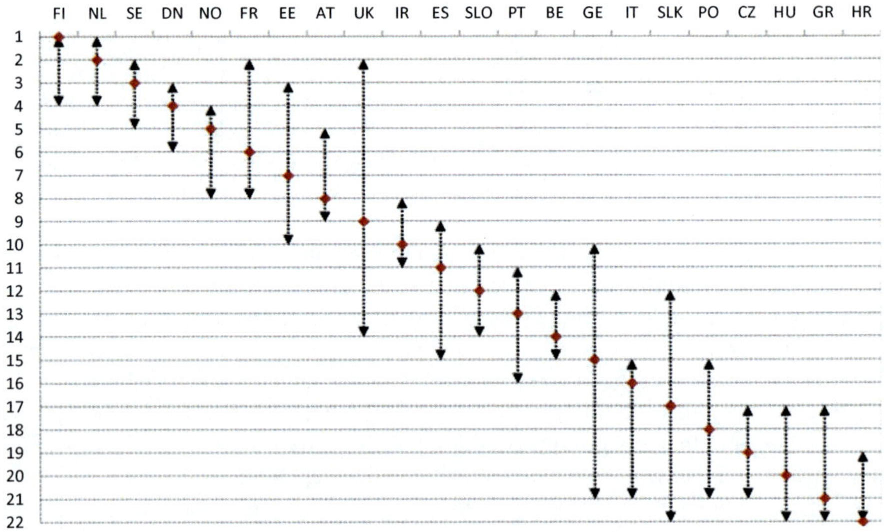

Fig. 5 Extreme ranking of European countries—Phase B

6.2 *Phase C—25 Reference Countries*

Incorporation of new countries in the reference ranking

In Phase C, the analyst creates five additional reference countries (C21–C25). In doing so, he makes sure to treat appropriately the points in the criteria in which the corresponding marginal value functions showed the greatest instability. The DM, from his side, inserted again the new reference countries into the pre-existing ranking, without changing the relevant positions of the former reference countries (Table 4).

Table 4 Integrated ranking of the 25 reference countries

Ranking positions	Fictitious countries	g_1	g_2	g_3	g_4	g_5	g_6	g_7	g_8
1	C6	85	75	2.5	85	0.8	40	80	60
2	C8	85	75	3.0	65	0.6	50	80	60
3	C7	85	75	2.5	65	0.6	50	85	60
4	C20	95	75	1.0	95	1.0	20	60	75
5	C17	85	65	2.0	95	0.8	40	80	60
6	C5	95	70	2.0	85	0.3	40	80	60
7	C19	80	60	3.0	75	0.8	60	80	70
8	C12	90	70	1.0	70	0.3	40	85	75
9	C1	85	70	2.0	75	0.6	40	80	60
10	C13	85	60	2.0	75	0.8	50	80	60
11	C14	80	60	2.0	80	0.8	65	80	60
12	C2	95	60	2.0	75	0.6	40	80	60
13	C4	95	70	2.0	55	0.4	40	80	60
14	**C22**	**75**	**70**	**3.0**	**80**	**0.7**	**40**	**90**	**20**
15	C9	85	75	3.0	65	0.6	30	70	50
16	C3	85	60	2.5	75	0.6	40	80	60
17	C18	90	70	1.0	75	0.6	40	70	50
18	**C24**	**85**	**50**	**0.5**	**90**	**0.7**	**65**	**85**	**80**
19	C10	80	65	3.0	65	0.6	40	80	40
20	C16	85	65	2.5	80	0.4	40	60	40
21	**C23**	**90**	**60**	**1.0**	**80**	**0.4**	**60**	**60**	**50**
22	**C25**	**60**	**45**	**4.0**	**50**	**0.8**	**50**	**85**	**90**
23	C15	90	70	2.0	50	0.4	40	60	30
24	**C21**	**90**	**75**	**2.5**	**40**	**0.2**	**20**	**50**	**70**
25	C11	80	60	1.0	65	0.3	20	70	40

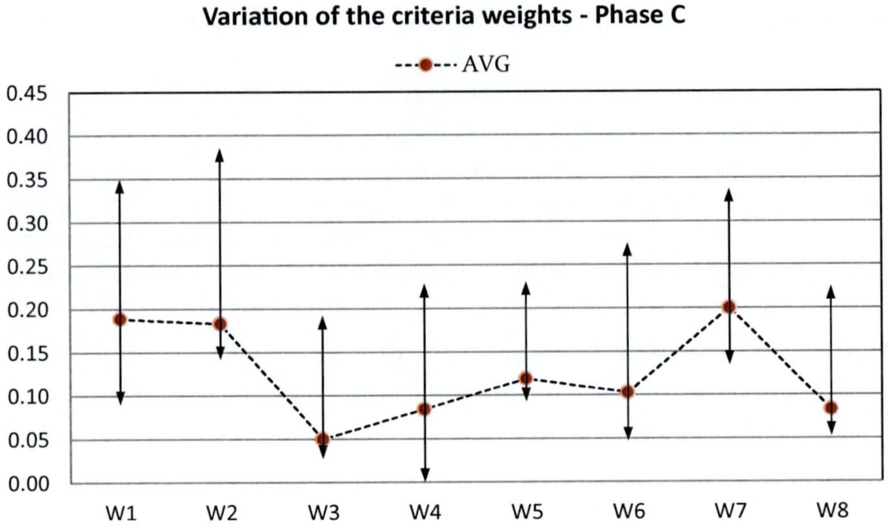

Fig. 6 Variation range of the criteria weights—Phase C

UTASTAR application and robustness control

The application of UTASTAR through the GAMS platform begins again with the validation that the sum of the errors is zero, before estimating the values of the model parameters. As in the previous phases, Fig. 6 shows the possible variation ranges and the average values of the weights of the criteria for all runs.

As expected, the relevant charts show a decrease in the possible range of both the marginal variables w_{ij} as well as the criteria weights W_i. This improvement is also reflected in the value of the *ASI*, which amounted to 0.971 compared to the value of 0.936 in the previous phase. This procedure continues flawlessly to the consideration of a representative model and to its application to the real data of the 22 European countries.

Robustness control of the country ranking

The formulation of a representative model makes it possible to calculate a new e-government ranking of the European countries. The ranking is obtained by recalculating the total value $u(\boldsymbol{g})$ of each country, using the most representative additive value model that resulted from this Phase. This ranking is shown in Table 5.

As in the previous phase, the extreme ranking analysis is implemented to achieve an overview of the stability of the countries ranking. The results obtained from the relevant algorithm are presented in Fig. 7. This analysis reveals that the range of variation of the 22 countries has decreased significantly compared to the previous phase. The *ARRI* index for this stage is 2.95, while the *RARR* index fell to 9.3%.

The analyst needs more information to assess the robustness of the results. For this reason, he uses a statistical tool of the aggregation pole, which allows him to

Table 5 Ranking of European counties based on the model inferred from the 25 reference countries

Ranking position	Country	Global value	Ranking position	Country	Global value
1	Netherlands	0.940	**12**	Slovenia	0.700
2	Finland	0.886	**13**	Belgium	0.684
3	Sweden	0.861	**14**	Portugal	0.681
4	France	0.855	**15**	Slovakia	0.631
5	Denmark	0.841	**16**	Germany	0.628
6	Un. Kingdom	0.831	**17**	Italy	0.622
7	Norway	0.816	**18**	Poland	0.586
8	Estonia	0.781	**19**	Czech Republic	0.572
9	Austria	0.773	**20**	Hungary	0.534
10	Ireland	0.740	**21**	Greece	0.534
11	Spain	0.704	**22**	Croatia	0.513

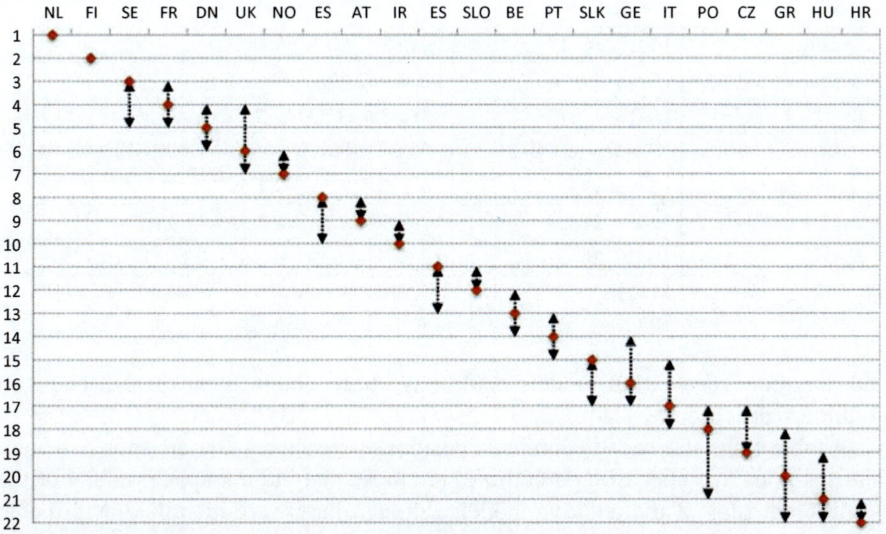

Fig. 7 Extreme ranking analysis of European countries—Phase C

control the outranking relations between the countries and their positions in the extreme ranking. For a sample of compatible additive value models, he calculates the probability that each country receives a specific position in the ranking. In this case, the models resulting from the 64 different runs during the application of the max – min process were selected. The statistical results are presented in Table 6 and show the possible ranking positions of each country and the frequency it achieves. Another technique to randomly generate weighting profiles from the feasible space

Table 6 Ranking position frequencies—Phase C

Country	Ranking position	Probability of occurrence
Netherlands	**1**	**100.00%**
Finland	**2**	**100.00%**
Sweden	**3**–4–5	**81.54%**–7.69%–10.77%
France	3–**4**–5	18.46%–**67.69%**–13.85%
Denmark	4–**5**–6	13.85%–**56.92%**–29.23%
Un. Kingdom	4–5–**6**–7	10.77%–18.46%–**67.69%**–3.08%
Norway	6–**7**	3.08%–**96.92%**
Austria	8–**9**	18.46%–**81.54%**
Estonia	**8**–9–10	**81.54%**–18.46%
Ireland	(9)–**10**	0.00%–**100.00%**
Slovenia	11–**12**	21.54%–**78.46%**
Spain	**11**–12–**13**	**78.46%**–9.23%
Belgium	11–12–**13**	12.31%–20%–**67.69%**
Portugal	(13)–**14**–15	0.00%–**67.69%**–32.31%
Germany	14–15–16–**17**	24.62%–20%–**55.38%**
Slovakia	15–**16**–17	20%–**67.69%**–12.31%
Italy	**15**–16–17–18	**55.38%**–12.31%–26.15%–6.15%
Poland	(17)–**18**–19–20–21	0.00%–**78.46%**–6.15%–3.08%–12.31%
Czech Republic	17–18–**19**	6.15%–12.31%–**81.54%**
Hungary	19–20–**21**–22	7.69%–38.46%–**52.31%**–1.54%
Greece	18–19–**20**–21–22	3.08%–4.62%–**58.46%**–23.08%–10.77%
Croatia	21–**22**	12.31%–**87.69%**

and, therefore, evaluation models could have been obtained by the Hit-and-Run algorithm, proposed by [15].

In Table 6, the statistically dominant positions for each country are shown in bold. The positions that appear in parentheses are those that did not appear in any of the 64 different runs of the additive model, although they were identified as possible positions by the extreme ranking analysis. This fact does not indicate a failure of the method, but it means that the specific weighting profiles that were considered, could not result in the specific positions.

All these robustness assessment results, as part of the tools of the aggregation pole, enable the analyst to decide on the level of e-government evaluation robustness. In this case, he considers that the robustness of the results, although quite satisfactory, can be further improved. For this reason, the analyst decides to repeat the whole process by returning to the disaggregation pole and executing one more round of robustness assessment, as defined by the bipolar robustness control methodology.

6.3 *Phase D—30 Reference Countries*

As mentioned above, the return to the disaggregation pole presupposes the extraction of additional information by the DM, so that they can be integrated into the existing model and contribute to the development of a more robust preference model.

Incorporation of new countries in the reference ranking

In full accordance with the algorithmic process of the previous phases, five new reference countries (C26–C30) are created in such a way as to compensate for the most unstable values in the w_{ij} parameters. The ranking given by DM, after the integration of the new reference countries in the ranking of phase C, is presented in Table 7.

Application of UTASTAR method and robustness control

The UTASTAR algorithm is again implemented similarly to the previous phases, by executing in GAMS the 64 runs that result in the range of the marginal variables w_{ij} and the criteria weights W_i. The variation ranges of the criteria weights, based on the preferential data of the 30 reference countries, are presented in Fig. 8. Similarly, the marginal value functions and their variation ranges are shown in Fig. 9.

Model robustness control

It is now clear from the above diagrams that the variation ranges, of both the marginal variables and the weights, have been significantly reduced, compared to the results of the previous phases. The improvement in the results is also reflected in the *ASI* index, which increased to 0.979. This allows the analyst to export a solid, fully representative model and proceed with its application to the real country data.

Robustness control of the country ranking

Having obtained a representative model, as the average weighting profile of the min–max procedure, the analyst safely applies it to the real data to achieve the ranking of European countries. The ranking is achieved by recalculating the global value $u(\mathbf{g})$ of each country (see Table 8).

As in the previous phase, the extreme ranking analysis gives a clear picture of the stability of the ranking. As shown in Fig. 10, there is again a significant reduction in the number of possible ranking positions for each country. The *ARRI* index for this phase decreases to 2.45, while the *RARR* index takes the value of 6.9%. This means that the average country in the ranking can occupy only 2.45 possible places in the ranking.

A statistical analysis is expected to confirm the robustness of the results and provide more information to the analyst and the DM, regarding the predominant positions of each country in the ranking. Using data on the w_{ij} values, deriving from the 64 different max–min runs, and an equal number of implementations of the additive value model, a statistical overview of the ranking frequencies is achieved. The results of the statistical analysis are shown in Table 9.

Table 7 Integrated ranking of the 30 reference countries

Ranking position	Fictitious country	g_1	g_2	g_3	g_4	g_5	g_6	g_7	g_8
1	**C28**	**95**	**85**	**3.5**	**75**	**0.8**	**40**	**80**	**75**
2	**C27**	**100**	**80**	**3.0**	**75**	**0.4**	**65**	**80**	**70**
3	**C29**	**80**	**70**	**4.0**	**75**	**0.6**	**65**	**90**	**85**
4	C6	85	75	2.5	85	0.8	40	80	60
5	C8	85	75	3.0	65	0.6	50	80	60
6	C7	85	75	2.5	65	0.6	50	85	60
7	C20	95	75	1.0	95	1.0	20	60	75
8	C17	85	65	2.0	95	0.8	40	80	60
9	C5	95	70	2.0	85	0.3	40	80	60
10	C19	80	60	3.0	75	0.8	60	80	70
11	C12	90	70	1.0	70	0.3	40	85	75
12	C1	85	70	2.0	75	0.6	40	80	60
13	C13	85	60	2.0	75	0.8	50	80	60
14	C14	80	60	2.0	80	0.8	65	80	60
15	**C30**	**85**	**65**	**3.5**	**95**	**0.8**	**35**	**70**	**50**
16	**C26**	**85**	**70**	**2.0**	**95**	**0.4**	**40**	**90**	**60**
17	C2	95	60	2.0	75	0.6	40	80	60
18	C4	95	70	2.0	55	0.4	40	80	60
19	C22	75	70	3.0	80	0.7	40	90	20
20	C9	85	75	3.0	65	0.6	30	70	50
21	C3	85	60	2.5	75	0.6	40	80	60
22	C18	90	70	1.0	75	0.6	40	70	50
23	C24	85	50	0.5	90	0.7	65	85	80
24	C10	80	65	3.0	65	0.6	40	80	40
25	C16	85	65	2.5	80	0.4	40	60	40
26	C23	90	60	1.0	80	0.4	60	60	50
27	C25	60	45	4.0	50	0.8	50	85	90
28	C15	90	70	2.0	50	0.4	40	60	30
29	C21	90	75	2.5	40	0.2	20	50	70
30	C11	80	60	1.0	65	0.3	20	70	40

The results of the statistical analysis reinforce the robustness assessment results obtained above, enriching them with valuable statistical information. At this stage, the final ranking of the 22 European countries, on which the analyst and the DM will end up, can be derived either from the results of the representative model (Table 8) or from the dominant positions of each country, as evidenced by the statistical analysis (Table 9).

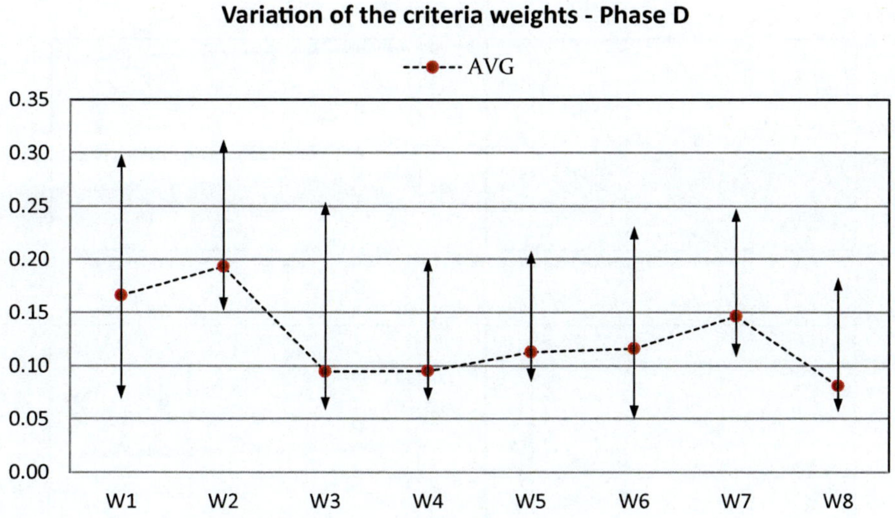

Fig. 8 Variation range of the criteria weights—Phase D

At the end of Phase D, and following a dialogue between the analyst and the DM, it is agreed that the results of the bipolar robustness control are sufficiently satisfactory and can confidently support the final evaluation of the 22 European countries. Based on the above, it is decided to terminate the robustness control process and consider the representative ranking as the final ranking of European e-government readiness.

The evolution of the robustness control indicators along with their gradual improvement, throughout the four-phase procedure that was followed, is presented in Table 10.

7 Conclusions

The robustness control methodology, presented in this paper, expands and systematizes robustness analysis as part of the ordinal regression multicriteria methods. The methodology consists of a strict, systematic, and interactive bipolar process, based on successive measurements and assessments of robustness. In particular, the two poles of the algorithm, namely, the disaggregation and the aggregation pole, interact and provide feedback to each other. Then, a set of robustness assessment measures and indicators are proposed, which are integrated into the two poles.

The methodology is successfully illustrated in the evaluation of e-government performance in Europe. The current model of e-government evaluation, despite its theoretical and technical soundness, presents proven instability, due to the way it is constructed, through the ordinal regression method UTASTAR. Specifically, the implicit way that the preferential parameters are extracted by the DM results in a

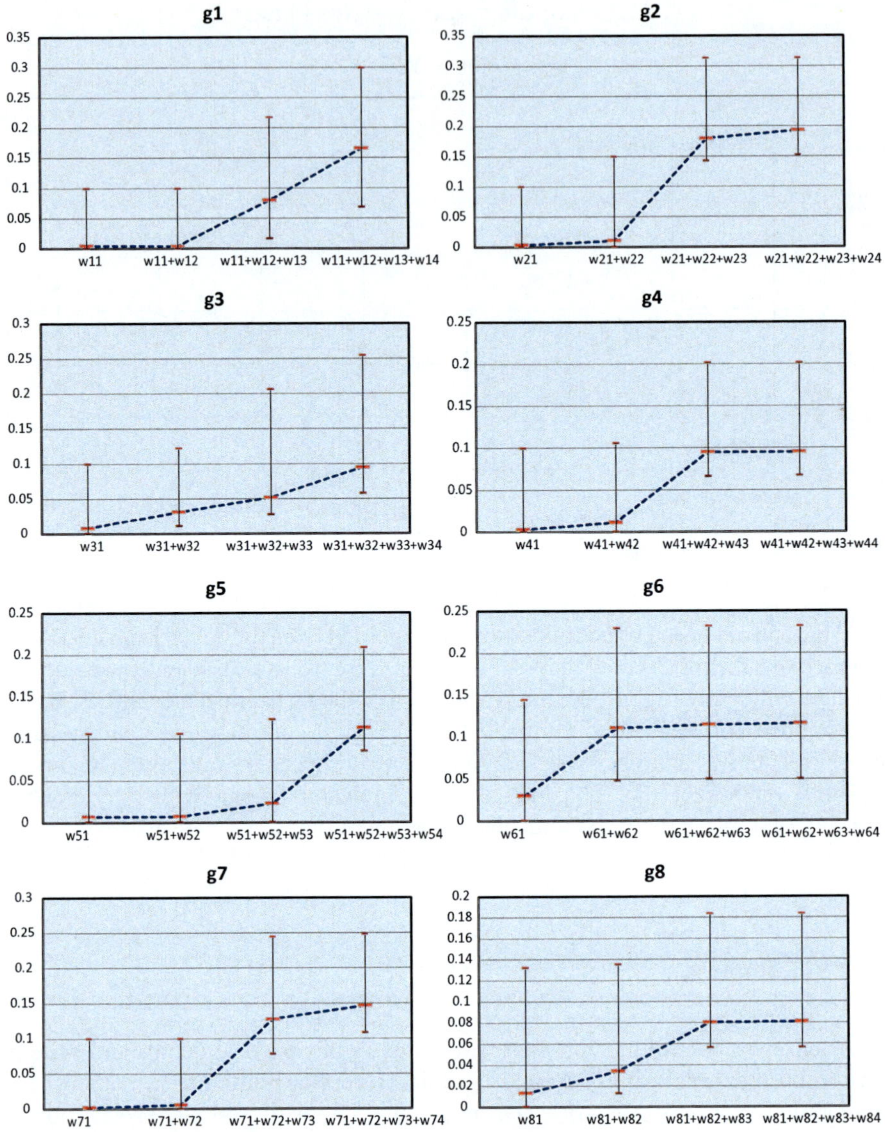

Fig. 9 Marginal value functions of the 30-reference country model

variety of different compatible mathematical models. Therefore, it is necessary to analyze the robustness of the model under development and its results.

However, as can be drawn from the implementation of the robustness control methodology, certain limitations hinder its universal application to other MCDA problems, with the most notable being the need for the DM's input throughout the whole procedure. When it comes to cases where the DM can only be reached once

Table 8 Ranking of European countries based on the model inferred from the 30 reference countries

Ranking position	Country	Global value	Ranking position	Country	Global value
1	Netherlands	0.881	**12**	Spain	0.615
2	Finland	0.828	**13**	Belgium	0.610
3	Sweden	0.807	**14**	Portugal	0.585
4	France	0.792	**15**	Germany	0.565
5	Denmark	0.785	**16**	Slovakia	0.532
6	Un. Kingdom	0.749	**17**	Italy	0.529
7	Norway	0.743	**18**	Poland	0.499
8	Austria	0.693	**19**	Czech Republic	0.483
9	Estonia	0.687	**20**	Hungary	0.437
10	Ireland	0.661	**21**	Croatia	0.411
11	Slovenia	0.629	**22**	Greece	0.409

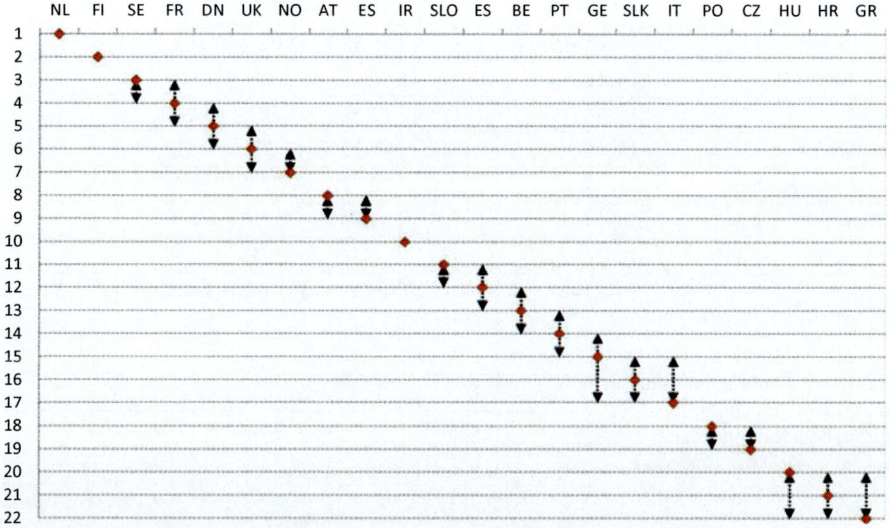

Fig. 10 Extreme ranking analysis of European countries—Phase D

or the arrangement of a meeting between the analyst and him/her is difficult, the methodology cannot be applied. In addition, the finalization of the procedure can be quite tiring for both the DM, who is required to provide preferential information consecutively, and for the analyst, who needs to perform several different analyses, some of which are proven demanding.

An important observation regarding the actual implementation of the methodology and the evolution of the robustness indicators is the non-relative and parallel

Table 9 Ranking position frequencies—Phase D

Country	Possible position	Probability of occurrence
Netherlands	**1**	**100.00%**
Finland	**2**	**100.00%**
Sweden	**3**–4	**93.85%**–6.15%
France	3–**4**–5	6.15%–**78.46%**–15.38%
Denmark	4–**5**–6	15,38%–**81.54%**–3.08%
United Kingdom	5–**6**–7	3.08%–**80%**–16.92%
Norway	6–**7**	16.92%–**83.08%**
Austria	**8**–9	**78.46%**–21.54%
Estonia	8–**9**	21.54%–**78.46%**
Ireland	**10**	**100.00%**
Slovenia	**11**–12	**80%**–20%
Spain	11–**12**–13	20%–**43.08%**–36.92%
Belgium	12–**13**–14	36.92%–**47.69%**–15.38%
Portugal	13–**14**–15	15.38%–**49.23%**–35.38%
Germany	14–**15**–16–17	35.38%–**43.08%**–6.15%–15.38%
Slovakia	15–**16**–17	9.23%–**49.23%**–41.54%
Italy	15–**16**–17	12.31%–**44.62%**–43.08%
Poland	**18**–19	**89.23%**–10.77%
Czech Republic	18–**19**	10.77%–**89.23%**
Hungary	**20**–21–22	**81.54%**–15.38%–3.08%
Croatia	20–**21**–22	1.54%–**66.15%**–32.31%
Greece	20–21–**22**	16.92%–18.46%–**64.62%**

Table 10 Evolution of the robustness indicators throughout the four-phase robustness control procedure

Phase	*ASI*	*ARRI*	*RARR*
A	0.885	–	–
B	0.936 (+5.8%)	6.27	24%
C	0.971 (+3.7%)	2.95 (–53.0%)	9.3% (–61.3%)
D	0.979 (+0.8%)	2.45 (–17.0%)	6.9% (–25.8%)

improvement of the indicators of the disaggregation and aggregation pole. In particular, it is clear that a small increase in the *ASI* index, which focuses on the variance of the model's parameters, causes significant improvements in the ARRI and RARR indicators, which assess the robustness of the results. This fact proves: (a) the optimism of the *ASI* index and more generally the indicators related to the disaggregation pole, and (b) the importance of measuring and analyzing robustness both during the model development phase but also in terms of the model's results. Indeed, ignorant of

(b) and based on the high values obtained by the robustness measures of the disaggregation pole (especially from Phase B onwards), we could arbitrarily avoid measuring the robustness of our results, convinced that it would be just as high and satisfactory. However, this, as evidenced by the algorithmic process we followed above, would have significant consequences to the reliability of the results, which were eventually not acceptable in the previous phases.

On the other hand, based on the above conclusion, one could argue that the analysis of robustness in the disaggregation pole is unnecessary, and has value only when performed in the aggregation pole. We can refute this claim, as demonstrated empirically in the application of the algorithm, and based on the fact that robust models produce largely robust rankings. Indeed, our experience of applying the algorithm to both poles has shown that analysis of the robustness of the model has often prevented us from conducting unnecessary and time-consuming analyses of the results, in cases when the model exhibited low robustness.

The prospects formed by this research effort are multidimensional and very interesting. In particular, it is considered valuable to extend the bipolar robustness control methodology, by introducing specific procedures that will address all cases of decision models and all possible applications and types of decision-makers. It is then necessary to examine the methodology in different types of decision models, using alternative multidisciplinary methods for individual decision problems, in order to obtain valuable feedback that will lay the foundations for further improvement. The ultimate goal of this research is to propose a holistic robustness control methodology that can be applied deliberately, providing both the analyst and the decision-maker with the opportunity to follow multiple alternative practices, when targeting the validation of the decision model in terms of each robustness.

Appendix A: Multicriteria Evaluation of European Countries on the Eight e-Government Evaluation Criteria

Country	g_1	g_2	g_3	g_4	g_5	g_6	g_7	g_8
Belgium	90.00	67.75	2.28	63.00	0.25	36.00	74.33	58.00
Czech Republic	88.00	63.76	1.91	56.00	0.25	21.33	87.67	44.50
Denmark	96.00	72.78	3.05	85.00	0.55	65.33	89.33	65.00
Germany	93.50	71.26	2.94	67.00	0.70	33.33	58.67	45.50
Estonia	89.50	69.50	1.74	87.00	0.76	35.00	80.00	77.50
Ireland	90.00	66.17	1.58	87.00	0.65	43.00	88.00	62.00
Greece	77.50	59.62	0.78	46.00	0.80	27.67	78.33	41.00
Spain	86.00	66.42	1.24	91.00	0.78	36.33	69.00	72.50
France	91.00	68.51	2.23	75.00	0.96	44.00	89.00	68.50

(continued)

(continued)

Country	g_1	g_2	g_3	g_4	g_5	g_6	g_7	g_8
Croatia	82.00	61.67	0.81	53.00	0.33	19.33	81.00	48.00
Italy	85.50	63.35	1.25	77.00	0.78	15.67	69.67	60.50
Hungary	81.50	61.96	1.41	45.00	0.45	34.33	82.33	35.50
Netherlands	98.00	77.40	1.98	82.00	1.00	57.67	80.67	65.50
Austria	89.50	67.41	2.81	86.00	0.63	40.33	80.67	70.50
Poland	84.00	62.50	0.87	76.00	0.49	17.33	81.67	51.00
Portugal	81.00	63.63	1.36	96.00	0.65	30.67	81.00	74.00
Slovenia	87.50	67.39	2.59	68.00	0.39	37.00	85.00	63.00
Slovakia	88.00	63.53	0.83	72.00	0.63	33.00	80.67	30.00
Finland	95.00	75.46	3.32	86.00	0.71	64.00	91.33	71.00
Sweden	94.00	74.94	3.21	83.00	0.61	60.33	90.67	68.50
Norway	95.00	71.42	1.69	78.00	0.69	64.33	84.33	63.50
Un. Kingdom	92.50	72.33	1.63	74.00	0.96	35.00	75.00	51.00

Appendix B: The UTASTAR Disaggregation Algorithm

The UTASTAR method is an improved version of the original UTA method (see [9]). UTASTAR uses a double positive error function, so that the value of each reference action $a \in A_R$ can be written as:

$$u'[\mathbf{g}(a)] = \sum_{i=1}^{n} u_i[g_i(a)] - \sigma^+(a) + \sigma^-(a)\ \forall a \in A_R$$

where σ^+ and σ^- are the underestimation and the overestimation error, respectively.

Based on the above, the UTASTAR algorithm may be summarized in the following steps:

Step 1.

Express the global value of reference actions $u\big[\mathbf{g}(a_k)\big], k = 1, 2, \ldots m$, first in terms of the marginal values $u_i(g_i)$, and then in terms of the variables w_{ij}:

$$\begin{cases} u_i(g_i^1) = 0 \ \ \forall i = 1, 2, \ldots, n \\ u_i(g_i^j) = \sum_{t=1}^{j-1} w_{it} \ \ \forall i = 1, 2, \ldots, n \text{ and } j = 2, 3, \ldots, \alpha_i \end{cases}$$

Step 2.

Introduce two error functions σ^+ and σ^- on A_R by writing for each pair of consecutive actions in the ranking of the analytic expressions:

$$\Delta(a_k, a_{k+1}) = u[\mathbf{g}(a_k)] - \sigma^+(a_k) + \sigma^-(a_k) - u[\mathbf{g}(a_{k+1})] + \sigma^+(a_{k+1}) - \sigma^-(a_{k+1})$$

Step 3.

Solve the following LP:

$$[\min]z = \sum_{k=1}^{m} [\sigma^+(a_k) + \sigma^-(a_k)]$$

Subject to:

$$\left.\begin{array}{l}\Delta(a_k, a_{k+1}) \geq \delta \text{ if } a_k \succ a_{k+1} \\ \Delta(a_k, a_{k+1}) = 0 \text{ if } a_k \sim a_{k+1}\end{array}\right\} \forall k$$

$$\sum_{i=1}^{n} \sum_{j=1}^{\alpha_i - 1} w_{ij} = 1$$

$$w_{ij} \geq 0,\ \sigma^+(a_k) \geq 0,\ \ \sigma^-(a_k) \geq 0\ \ \forall i, j \text{ and } k$$

where a_k and a_{k+1} are two successive actions in the DM's ranking and δ is a small positive number, indicating the preference threshold between the two actions.

Step 4.

Test the existence of multiple or near optimal solutions of the LP (stability/robustness analysis); in case of non-uniqueness, find the mean additive value function as the most representative (barycenter) of those (near) optimal solutions which maximize/minimize the objective functions:

$$u_i(g_i^j) = \sum_{t=1}^{j-1} w_{it} \text{ for } i = 1, 2, \ldots, n \text{ and } j = 2, 3, \ldots, \alpha_i.$$

on the polyhedron of the constraints of the previous LP bounded by the new constraint:

$$\sum_{k=1}^{m} \left[\sigma^+(a_k) + \sigma^-(a_k)\right] \leq z^* + \varepsilon$$

where z^* is the optimal value of the LP in step 3 and ε is a very small positive number.

The number of LPs that have to be solved in this step (and the corresponding value functions obtained) is $2 \cdot \sum_{i=1}^{n} (\alpha_i - 1)$. In most of the UTASTAR applications, one usually seeks value functions that are free of errors (all error variables σ are zero), since no relaxation from the minimal error is allowed ($\varepsilon = 0$).

References

1. Abramson, A.M., Means, E.G.: *E-Government*, Price Waterhouse Coopers, Endowment for the Business of Government, Rowman & Littlefield Publishers Inc. (2001)
2. Dawes, S.S.: The evolution and continuing challenges of e-governance. Public Admin. Rev. **68**(6), 82–102 (2008)
3. Dyer, J.S.: Multiattribute utility theory (MAUT). In: Greco, S., Ehrgott, M., Figueira, J. (eds.) Multiple Criteria Analysis: State of the Art Surveys, vol. 1, 2nd edn., pp. 285–314. Springer, New York (2016)
4. European Commission: *eGovernment benchmark 2019: Empowering Europeans through trusted digital public services*, European Union (2019)
5. Greco, S., Słowiński, R., Figueira, J., Mousseau, V.: *Robust ordinal regression*, Chapter 8, In: Ehrgott, M., Greco, S., Figueira, J. (eds.) Trends in Multiple Criteria Decision Analysis. Springer, Berlin (2010)
6. Grigoroudis, E., Siskos, Y.: Customer Satisfaction Evaluation: Methods for Measuring and Implementing Service Quality. Springer, New York (2010)
7. Heeks, R., Bailur, S.: Analyzing e-government research: Perspectives, philosophies, theories, methods, and practice. Govern. Inf. Q. **24**, 243–265 (2007)
8. Jacquet-Lagrèze, E., Siskos, Y.: Assessing a set of additive utility functions for multicriteria decision making: The UTA method. Eur. J. Oper. Res. **10**(2), 151–164 (1982)
9. Jacquet-Lagrèze, E., Siskos, Y.: Preference disaggregation: 20 years of MCDA experience. Eur. J. Oper. Res. **130**, 233–245 (2001)
10. Kadzinski, M., Greco, S., Slowinski, R.: Extreme ranking analysis in robust ordinal regression. Omega **40**, 488–501 (2012)
11. Kadzinski, M., Greco, S., Slowinski, R.: Selection of a representative value function in robust multiple criteria ranking and choice. Eur. J. Oper. Res. **217**, 541–553 (2012)
12. Matsatsinis, N.F., Grigoroudis, E. (eds.): Preference Disaggregation in Multiple Criteria Decision Analysis: Essays in Honor of Yannis Siskos. Springer, New York (2018)
13. Siskos, E., Askounis, D., Psarras, J.: Multicriteria decision support for global e-government evaluation. Omega **46**, 51–63 (2014)
14. Siskos, Y., Grigoroudis, E., Matsatsinis, N.F.: UTA Methods. In: Greco, S., Ehrgott, M., Figueira, J. (eds.) Multiple Criteria Analysis: State of the Art Surveys, vol. 1, 2nd edn., pp. 315–362. Springer, New York (2016)
15. Tervonen, T., Valkenhoef, G.V., Baştürk, N., Postmus, D.: Hit-And-Run enables efficient weight generation for simulation-based multiple criteria decision analysis. Eur. J. Oper. Res. **224**, 552–559 (2012)
16. United Nations: UN E-government survey 2020. New York (2020)

The COMET Method: Study Case of Swimming Training Progress

Jakub Więckowski and Jarosław Watróbski

Abstract Many factors influence the preparation of the top form for the main swimming competition. The direct preparation period lasts for about 3 months, during which the competitor swims hundreds of kilometers on different intensities. Heart rate measurement after the tasks can be used to assess the progress of preparations and determine the trend in which the athlete's form is heading. In this paper, we collected real data from three professional swimmers. Using the multi-criteria method called COMET, the values of attributes characteristic for each swimmer were introduced. The results of the preference values obtained from the COMET method were used in the linear regression to determine the trend of the player's form. The obtained model allows a broader analysis of the progress in terms of particular criteria sensitivity and robustness analyses.

Abbreviations

MCDA	Multi-Criteria Decision Analysis
COMET	Characteristic Objects Method
PROMETHEE	Preference Ranking Organization Method for Enrichment of Evaluations
ELECTRE	Elimination Et Choice Translating Reality
AHP	Analytic Hierarchy Process
ANP	Analytic Network Process
COs	Characteristic Objects
TFN	Triangular Fuzzy Number
mmol	Millimole

J. Więckowski
West Pomeranian University of Technology in Szczecin,
ul. Żołnierska 49, 71-210 Szczecin, Poland
e-mail: jakub-wieckowski@zut.edu.pl

J. Watróbski (✉)
University of Szczecin, Mickiewicza 64, 71-101 Szczecin, Poland
e-mail: jaroslaw.watrobski@usz.edu.pl

A. J. Kulkarni (ed.), *Multiple Criteria Decision Making*, Studies in Systems, Decision and Control 407, https://doi.org/10.1007/978-981-16-7414-3_8

Mathematical Symbols

$\mu_{\tilde{A}}$	The value of the membership function
C_r	Set of the fuzzy numbers, where C_i represents each criterion
CO	Characteristic Objects as Cartesian Product
a_{ij}	Element of the matrix of expert judgement, where $i = 1, 2, \ldots, n$ is and index of row in matrix and $j = 1, 2, \ldots, m$ is an index of matrix' column
SJ_i	The sum of the i-th row of matrix of expert judgement
y	The value of linear function
β_0	The coefficient of the independent variable in linear function
β_1	A directional coefficient in linear function
x	The independent variable in linear function

1 Introduction

Professional athletes reach the heights of their abilities at every training session, hoping that this will lead them to the desired success. After consulting with the trainer and determining the volume and intensity of training, their task is to shoot as accurately as possible at the previously set level of lactation [30, 47, 50]. A swimmer who has swum hundreds of kilometers in the pool in his career can feel the oxygen threshold at which he is making an effort. However, sometimes this feeling is not enough to evaluate fatigue and the realization of training pressures. It is where the heart rate measurement comes in. On its basis, it is possible to determine what intensity swimmers have reached during a given effort. Moreover, the heart rate helps the trainer and the competitor assess how the competitor reacts to training impulses, whether his form in preparation for the competition increases or whether the competitor becomes tired [22, 32]. A reliable determinant of this factor is the use of the technique based on heart rate measurement during and after the main training task. A swimmer puts his point and middle finger on the carotid artery or radial artery in the interval between the successive lengths covered by the swimmer and counts the number of heartbeats for 10 s [16, 19]. Each athlete knows how many beats correspond to the respective intensity. However, this is an individual issue, and the thresholds may differ from person to person. For example, the generally accepted number of heartbeats with an intensity of 4 mmol (millimole) of lactic acid is 26 beats [13, 17].

Afterward, when the contestant has completed the whole task, he takes further measurements. This time it is as follows: after 5 s from the end of the task, the competitor puts two fingers on the designated places and counts the number of heartbeats for 10 s, waits 20 s, then another 10 s measurement, another 20 s break, and another 10 s of heartbeat counting. This technique will allow you to evaluate how the body reacts to the task or how quickly it can eliminate fatigue. In a demanding training period, the heart rate decrease is not significant, but the closer to the main start, the greater the differences in subsequent measurements should be [2, 49]. For example, the heart rate distribution at the level of 30, 27, 24 beats shows that the competitor has completed the task at the intensity level of 10–12 mmol and his body is managing well to eliminate fatigue, but this is not yet the optimal form, in which he will achieve the best results. For example, heart rates of 30, 26, 18 beats show that the athlete is approaching the peak form, as indicated by a significant drop in the third measurement. The heart rate of a swimmer who is overtrained will be characterized by a slight difference between the next measurements, for example, 31, 29, 28 heart beats. In this case, the athlete reached a high training intensity of 12 mmol, and a slight decrease in heart rate indicates the inability to eliminate fatigue and lactic acid from the body immediately [8, 18, 31]. As we know, this situation is not favorable for a competitor who can take part in 2–5 events in one day at the competition. Fatigue build-up limits the speed of regeneration and lowers the threshold from which the competitor approaches the next start. Fatigue is not a desirable effect among athletes because it is associated with the need to reduce the intensity of training to allow the body to regenerate and gain strength for the next weeks of preparation for the competition [21]. On the other hand, during the most challenging period of preparation, it is good to train in such a way that the fatigue achieved is close to exhaustion. To achieve the highest results, it is often necessary to go beyond the mental and physical comfort zone [18]. Such a crossing of boundaries is the training performed near muscle fall. However, there is a thin line between sensible and controlled training that requires reaching the heights of ability and training that leads to long-term exhaustion of both the muscular and nervous systems [22]. As in everything, a golden mean must be found. In this case, a training plan will provide the player with an appropriate ratio of units at high intensity to those intended for recovery [12, 20].

Identifying a mathematical model for managing swimming training goes beyond the classic optimization paradigm [3, 7]. The multi-criteria nature and the context of the problem under consideration clearly show that the MCDA (Multi-Criteria Decision Analysis) methods appropriateness. MCDA methods allow us to solve problems where different aspects need to be analyzed to get the solution most satisfying the decision-maker [4, 57]. Thus, the optimization paradigm is replaced by a rational solution choice that guarantees meeting all the decision-maker objectives according to his preference system [39].

Despite a large number of MCDA methods, the correct and accurate mapping of measurement data and preferences of the decision-maker in the model remains a challenge [6, 45, 52]. For example, the modeling of the criteria performance may be performed in different MCDA methods in a different way using, e.g., quantitative weights (Electre [39] or Promethee [5] family methods) qualitative weights (Melchior [29], Oreste [33]) or relative weights of criteria (AHP/ANP family methods in the crisp and fuzzy form [28, 58]). Regardless of the type of weights adopted at the preference modeling stage, it is still difficult to reflect them correctly and transfer the appropriate numerical values consistent with the model being developed [6]. Another current research challenge is to properly reflect the imprecision of measurement data [37] and preferences of the decision-maker in the model [39].

Despite the great popularity of outranking relation-based MCDA methods (ELECTRE and Promethee family), where the uncertainty of preferences is described using thresholds, the currently emerging MCDA-based solutions are based on the use of fuzzy number arithmetic [11]. The latter proved to be a powerful tool when dealing with data and preferences uncertainty, and the newly developed MCDA methods are based on successive generations of fuzzy sets representing the great potential for fuzzy sets in the MCDA domain [10]. Regardless of the above, and due to the formal assumptions of the computational algorithms themselves, most of the MCDA methods from the so-called American school [52], have the undesirable effect of linear compensation of criteria. In many works, the so-called rank reversal problem [51] is raised.

In response to these shortcomings, the practical application of the COMET method is presented in the article. We propose a full domain model that allows us to evaluate the swimmer's form based on the heart rate measurement and the number of days to the main event. The COMET method has superb abilities in terms of a very limited linear compensation effect, and indirect defining weights for criteria in the problem as well [23]. Additionally, the final ranking is, at the same time, free of rank reversal problem [46]. Moreover, in combination with fuzzy logic, which allows us to introduce elements of uncertainty into the considered environment (in terms of data and DM preferences), it allows for a wide application in solving multi-criteria problems [24, 54–56]. The assumptions of the Fuzzy Sets theory allow the expert to define values in a blurred way, making it easier to analyze a problem with a large number of criteria [25, 27]. What is important at the stage of preference modeling, COMET additionally allows, on the basis of obtained expert values, to faithfully reflect the form of preference function for each criterion [26, 40, 42, 48].

The rest of the paper is organized as follows: the next Sect. 2 discusses learning curves, the theory of Triangular Fuzzy Numbers and the COMET method. Principles of the technique of linear regression are also briefly recalled. In Sect. 3, an experiment was performed on a previously built decision model. The study was based on the results of three athletes during the 2018 season. The measurements were taken over 70 days in preparation for the national event. Section 4 includes the presentation and discussion of the results, as well as the comparison of the results between the individual athletes. The last Sect. 5 presents a summary and conclusions from the study.

2 Fuzzy Logic—Preliminaries

Lofti Zadeh introduced the idea of fuzzy sets in his paper in 1965 [57]. The growing importance of the Fuzzy Set Theory in model creation in numerous scientific fields has proven to be an effective way to approach and solve multi-criteria decision problems [4, 14, 38]. The assumptions included in the fuzzy logic allowed for attempts to solve problems with uncertain characteristics. Expertise could be expressed with fuzzy values, which guaranteed satisfactory results despite the uncertainties that existed. The necessary concepts of the Fuzzy Set Theory are described as follows [9, 35]:

The fuzzy set and the membership function—the characteristic function μ_A of a crisp set $A \subseteq X$ assigns a value of either 0 or 1 to each member of X, as well as the crisp sets only allow a full membership ($\mu_A(x) = 1$) or no membership at all ($\mu_A(x) = 0$). This function can be generalized to a function $\mu_{\tilde{A}}$ so that the value assigned to the element of the universal set X falls within a specified range, i.e. $\mu_{\tilde{A}} : X \rightarrow [0, 1]$. The assigned value indicates the degree of membership of the element in the set A. The function $\mu_{\tilde{A}}$ is called a membership function and the set $\tilde{A} = (x, \mu_{\tilde{A}}(x))$, where $x \in X$, defined by $\mu_{\tilde{A}}(x)$ for each $x \in X$ is called a fuzzy set [59].

The triangular fuzzy number (TFN)—a fuzzy set $\tilde{A}$, defined on the universal set of real numbers $\Re$, is told to be a triangular fuzzy number $\tilde{A}(a, m, b)$ if its membership function has the following form (1):

$$\mu_{\tilde{A}}(x, a, m, b) = \begin{cases} 0 & x \leq a \\ \frac{x-a}{m-a} & a \leq x \leq m \\ 1 & x = m \\ \frac{b-x}{b-m} & m \leq x \leq b \\ 0 & x \geq b \end{cases} \tag{1}$$

and the following characteristics (2), (3):

$$x_1, x_2 \in [a, b] \wedge x_2 > x_1 \Rightarrow \mu_{\tilde{A}}(x_2) > \mu_{\tilde{A}}(x_1) \tag{2}$$

$$x_1, x_2 \in [b, c] \wedge x_2 > x_1 \Rightarrow \mu_{\tilde{A}}(x_2) > \mu_{\tilde{A}}(x_1) \tag{3}$$

The support of a TFN—the support of a TFN $\tilde{A}$ is defined as a crisp subset of the $\tilde{A}$ set in which all elements have a non-zero membership value in the $\tilde{A}$ set (4):

$$S(\tilde{A}) = \{x : \mu_{\tilde{A}}(x) > 0\} = [a, b] \tag{4}$$

The core of a TFN—the core of a TFN $\tilde{A}$ is a singleton (one-element fuzzy set) with the membership value equal to 1 (5):

$$C(\tilde{A}) = \{x : \mu_{\tilde{A}}(x) = 1\} = m \tag{5}$$

The fuzzy rule—the single fuzzy rule can be based on the Modus Ponens tautology [34]. The reasoning process uses the $IF - THEN$, OR, and AND logical connectives.

The rule base—the rule base consists of logical rules determining the causal relationships existing in the system between the input and output fuzzy sets [34].

The T-norm operator (product)—the T-norm operator is a T function modeling the AND intersection operation of two or more fuzzy numbers, e.g., $\tilde{A}$ and $\tilde{B}$. In this paper, only the ordinary product of real numbers is used as the T-norm operator [15] (6):

$$\mu_{\tilde{A}}(x) AND \mu_{\tilde{B}}(y) = \mu_{\tilde{A}}(x) \cdot \mu_{\tilde{B}}(y) \tag{6}$$

3 The Characteristic Objects Method

The COMET method belongs to the group of MCDA. It is entirely free of the rank reversal phenomenon. Formally, the comet method is based on the assumption of full comparability of alternatives, thus assuming the existence of two preference functions (equivalence and preferences) of the decision options. This implies that all decision variants' performance to compare concerning all criteria is expressed on the quantitative scale. The identification model is based on expert knowledge and Characteristic Objects (COs) comparison. Thus there is no need to determine weights for identified criteria directly. However, it is worth pointing out that this does not result in the loss of preference information—the COMET method takes into account the fact that preference functions do not have to be linear, which results in the weighting of criteria being indirectly transferred to the resultant model, thus retaining in the model better opportunities to reflect any form of preference function (Fig. 1).

Rank Reversal phenomenon is a big shortcoming of many MCDA methods. Adding or subtracting a defined group of alternatives may cause a change of position between the previous and current rankings. It is a phenomenon that does not occur in the COMET method. It is a method completely free of this phenomenon since a set of characteristic objects replaces the decision alternatives. These objects act as navigation buoys and allow us to determine the level of preferences of decision-making alternatives. Thus, the assessed and the metric space building set is separated here, and thanks to the Reversal ranking, it does not occur in the COMET method [46].

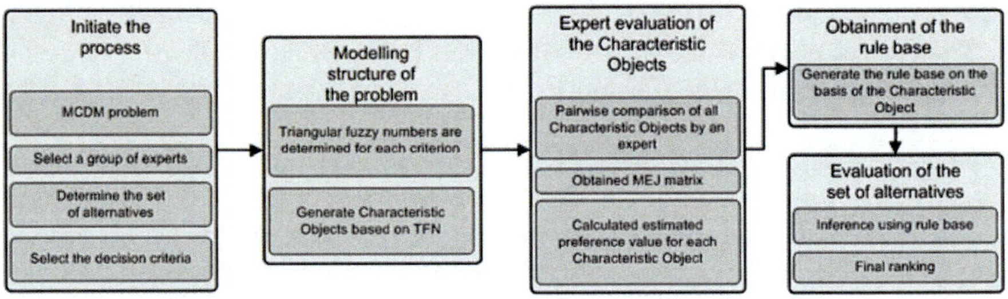

Fig. 1 The full identification procedure by using the COMET method

Basing COMET methods on the fuzzy numbers arithmetic makes this method a powerful tool because it addresses the natural uncertainty of data. It is worth adding that in previous works, the comet method's accuracy was verified [44], and COMET proved to be a useful tool to handle various forms of uncertainty and uncertainty of measurement data [10, 43]. The formal notation of the comet method should be shortly recalled [41, 44, 53].

Step 1. Define the space of the problem—the expert determines the dimensionality of the problem by selecting the number r of criteria, $C_1, C_2, \ldots, C_r$. Then, the set of fuzzy numbers for each criterion C_i is selected (7):

$$C_r = \{\tilde{C}_{r1}, \tilde{C}_{r2}, \ldots, \tilde{C}_{rc_r}\} \tag{7}$$

where $c_1, c_2, \ldots, c_r$ are numbers of the fuzzy numbers for all criteria.

Step 2. Generate characteristic objects—The characteristic objects (CO) are obtained by using the Cartesian Product of fuzzy numbers cores for each criteria as follows (8):

$$CO = C(C_1) \times C(C_2) \times \cdots \times C(C_r) \tag{8}$$

Step 3. Rank the characteristic objects—the expert determines the Matrix of Expert Judgment (MEJ). It is a result of pairwise comparison of the COs by the problem expert. The MEJ matrix contains results of comparing characteristic objects by the expert, where α_{ij} is the result of comparing CO_i and CO_j by the expert. The function f_{exp} denotes the mental function of the expert. It depends solely on the knowledge of the expert and can be presented as (9). Afterward, the vertical vector of the Summed Judgments (SJ) is obtained as follows (10):

$$\alpha_{ij} = \begin{cases} 0.0, & f_{exp}(CO_i) < f_{exp}(CO_j) \\ 0.5, & f_{exp}(CO_i) = f_{exp}(CO_j) \\ 1.0, & f_{exp}(CO_i) > f_{exp}(CO_j) \end{cases} \tag{9}$$

$$SJ_i = \sum_{j=1}^{t} \alpha_{ij} \tag{10}$$

Finally, values of preference are approximated for each characteristic object. As a result, the vertical vector P is obtained, where i-th row contains the approximate value of preference for CO_i.

Step 4. The rule base—each characteristic object and value of preference is converted to a fuzzy rule as follows (11):

$$IF\ C(\tilde{C}_{1i})\ AND\ C(\tilde{C}_{2i})\ AND\ \ldots\ THEN\ P_i \tag{11}$$

In this way, the complete fuzzy rule base is obtained.

Step 5. Inference and final ranking—each alternative is presented as a set of crisp numbers (e.g., $A_i = \{a_{1i}, a_{2i}, \ldots, a_{ri}\}$). This set corresponds to criteria $C_1, C_2, \ldots, C_r$. Mamdani's fuzzy inference method is used to compute preference of i-th alternative. The rule base guarantees that the obtained results are unequivocal.

3.1 Linear Regression

A matching regression line or curve represents the estimated expected value of the variable y at the specific values of another variable or variables x. In the simplest case, a constant or linear function is matched:

$$y = \beta_0 + \beta_1 \cdot x \tag{12}$$

Regression, in general, is a problem of conditionally expected value estimation. Linear regression is called linear because the assumed model of dependence between dependent and independent variables is a linear (affine) transformation of parameters, represented in a multidimensional case by a matrix. Directional coefficient β_1 informs us about an upward or downward trend of a given phenomenon. It informs us about the slope of the regression slope.

4 Study Case

The considered problem concerns the assessment of the form of three professional swimmers based on heart rate measurements. Depending on the intensity of the effort, the results of the heart rate measurement vary. Using the technique of measuring heartbeats three times in time interval after the task is completed, we can evaluate the work done by the competitor and the direction in which the swimmer's form is heading. The question is how the heart rate obtained by a competitor correlates with the number of days to the main start? The closer to the target event, the more significant the drop in the following values during the heart rate measurements should be (Fig. 2).

Many coaches use the heart rate measurement technique during and after the tasks. It is a reliable factor that helps to react appropriately and introduce individual variations into the training to maximize the effects and lead the athlete to top form at the final sports event [1, 36]. The model developed to evaluate the form of professional swimmers based on the heart rate analysis uses the COMET method due to its simplicity and flexibility. The obtained measurement data for three professional swimmers from particular days are presented in Table 1. The randomize data presenting values obtained by alternatives A_1–A_5 and A_6–A_{10} are presented in Tables 2 and 3. After 5 s from the end of the task, the number of heartbeats is counted for 10 s (it is C_1). Then, we wait 20 s, and pulses are counted one more again by 10 s (C_2),

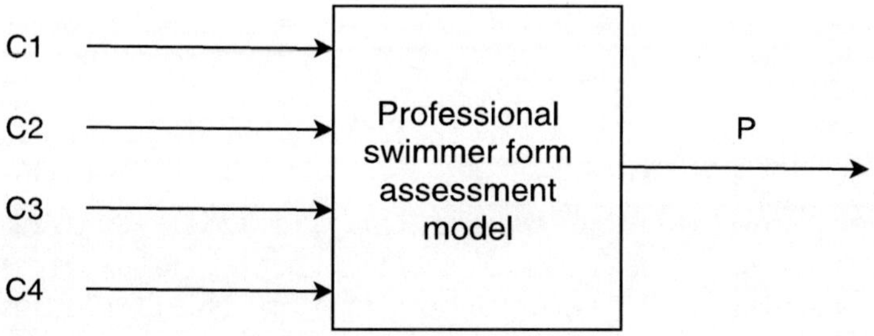

Fig. 2 Schematic of the professional swimmer assessment model

Table 1 Decision matrix presented heart rate measures of professional swimmers referred as $\mathbf{A}_1$–$\mathbf{A}_5$, for given criteria represented as $\mathbf{C}_1$–$\mathbf{C}_4$, and analyzed training units as $\mathbf{B}_i$

B	**A₁**				**A₂**				**A₃**			
$\mathbf{B_i}$	$\mathbf{C_1}$	$\mathbf{C_2}$	$\mathbf{C_3}$	$\mathbf{C_4}$	$\mathbf{C_1}$	$\mathbf{C_2}$	$\mathbf{C_3}$	$\mathbf{C_4}$	$\mathbf{C_1}$	$\mathbf{C_2}$	$\mathbf{C_3}$	$\mathbf{C_4}$
B_1	28	23	16	75	29	26	24	76	31	24	29	76
B_2	29	25	22	73	27	24	20	75	27	21	17	75
B_3	29	27	27	71	30	22	19	73	32	29	26	71
B_4	31	27	24	66	31	27	22	69	30	26	22	69
B_5	29	28	25	64	30	26	24	64	32	28	23	66
B_6	26	22	18	63	26	22	19	63	32	28	24	42
B_7	28	18	16	61	26	22	19	61	33	27	25	40
B_8	30	29	25	42	30	27	25	42	30	23	21	39
B_9	30	27	23	40	31	29	26	40	33	27	24	38
B_{10}	28	22	17	39	27	24	20	39	30	24	20	32
B_{11}	31	27	24	38	31	29	27	38	31	26	23	25
B_{12}	29	20	16	32	29	26	22	32	30	27	23	24
B_{13}	30	27	21	25	29	27	24	25	31	26	24	22
B_{14}	31	29	24	19	31	28	26	19	32	27	26	19
B_{15}	32	26	24	17	32	27	24	17	31	27	23	17
B_{16}	29	18	14	10	29	27	23	10	30	26	21	10

another 20 s break, and another 10 s of heartbeat counting (C_3). So criteria C_1 to C_3 means heart rate value, which was measure three times after the end of the task. The criterion C_4 means the number of days left to the event.

Basing on expert knowledge, four input components were defined—first, second, and third heart rate beats from the measurement and the number of days to the competition. Over 70 days, 16 training units were analyzed at intervals. The units that required considerable effort from the players to implement the planned training program were selected. Measurements were taken at training sessions with an intensity not less than that at which the athletes achieved an acidity level of 4 mmol. The most

Table 2 Randomized decision matrix for criteria $\mathbf{C}_1$–$\mathbf{C}_4$ for alternatives $\mathbf{A}_1$–$\mathbf{A}_5$, where $\mathbf{B}_i$ is analyzed training unit

$\mathbf{B_i}$	A_1				A_2				A_3				A_4				A_5			
	C_1	C_2	C_3	C_4	C_1	C_2	C_3	C_4	C_1	C_2	C_3	C_4	C_1	C_2	C_3	C_4	C_1	C_2	C_3	C_4
B_1	28	25	23	73	27	23	17	71	28	22	16	75	28	26	22	72	27	22	18	69
B_2	31	27	24	69	28	24	19	68	29	24	20	73	29	24	17	68	29	26	23	63
B_3	33	28	23	64	30	26	22	66	28	26	24	71	27	22	16	62	28	22	21	61
B_4	30	28	22	61	31	27	23	60	30	27	23	62	27	25	23	61	26	23	20	59
B_5	32	26	21	58	32	30	27	52	31	28	26	59	29	27	26	55	29	24	22	57
B_6	32	29	23	54	29	26	24	50	31	29	26	57	30	28	22	52	30	28	25	52
B_7	27	23	18	47	30	27	22	49	30	26	21	53	31	29	27	44	32	29	26	50
B_8	30	28	21	44	32	28	26	46	27	23	22	48	32	29	29	43	28	26	20	45
B_9	31	25	20	39	32	29	27	43	32	29	28	42	30	28	27	38	26	22	16	40
B_{10}	29	27	23	37	30	27	24	37	33	30	29	40	31	30	29	34	27	24	21	33
B_{11}	28	27	19	36	28	25	22	32	30	28	27	33	30	27	26	29	28	25	20	31
B_{12}	32	29	25	28	29	26	18	28	29	27	26	31	30	28	27	26	27	22	18	29
B_{13}	31	28	24	23	31	29	27	24	31	28	23	22	33	31	28	21	26	24	17	25
B_{14}	30	25	23	19	26	24	18	19	33	27	22	16	32	30	29	19	26	23	18	17
B_{15}	30	26	23	15	30	28	21	14	30	26	22	12	33	30	29	17	29	23	17	16
B_{16}	29	25	20	12	31	27	25	11	28	24	20	10	30	29	28	14	27	24	16	12

Table 3 Randomized decision matrix for criteria $\mathbf{C}_1$–$\mathbf{C}_4$ for alternatives $\mathbf{A}_6$–$\mathbf{A}_{10}$, where $\mathbf{B}_i$ is analyzed training unit

$\mathbf{B}_i$	A_6				A_7				A_8				A_9				A_{10}			
	C_1	C_2	C_3	C_4	C_1	C_2	C_3	C_4	C_1	C_2	C_3	C_4	C_1	C_2	C_3	C_4	C_1	C_2	C_3	C_4
B_1	28	25	23	74	28	26	22	75	28	25	22	72	28	26	23	72	27	25	23	75
B_2	29	24	20	68	28	24	19	71	28	23	17	62	30	27	24	70	29	27	26	72
B_3	30	27	22	62	31	27	24	61	30	27	24	58	29	26	21	61	30	28	26	70
B_4	31	28	26	54	30	27	26	55	31	28	26	52	28	24	22	58	31	27	25	66
B_5	31	27	24	50	29	26	23	53	29	26	24	50	26	22	18	53	29	26	24	60
B_6	30	28	27	49	30	28	26	48	27	26	22	46	32	25	20	49	28	23	22	58
B_7	29	27	25	48	32	29	28	41	30	29	28	41	33	28	18	43	32	29	27	52
B_8	32	29	27	38	31	29	28	39	31	28	24	39	29	25	21	40	31	28	26	49
B_9	33	29	27	33	29	26	21	37	33	30	28	38	30	28	23	38	32	29	28	38
B_{10}	31	28	26	29	30	26	24	36	29	26	24	33	31	28	24	36	30	27	25	33
B_{11}	30	28	26	25	31	27	21	35	30	22	21	29	30	26	23	34	29	26	24	31
B_{12}	29	27	26	21	33	28	26	32	29	27	26	22	33	27	21	27	31	29	26	30
B_{13}	30	28	25	18	27	25	24	29	28	26	25	21	31	22	20	21	32	28	25	26
B_{14}	31	26	23	14	30	28	26	25	32	30	29	20	30	21	18	16	30	27	24	22
B_{15}	30	27	24	11	29	26	22	16	31	29	27	17	27	22	19	14	31	29	26	21
B_{16}	29	23	18	10	27	25	18	12	33	30	29	13	26	18	16	10	29	26	23	15

demanding training period was for training units B_3–B_{11}, while the time when a decrease in individual measurements should be noticeable was for units B_{14}–B_{16}.

The model's output was defined as one value—the evaluation of the competitor's form in a given situation. Model evaluation is represented as a value in the range of 0–1, where 1 determines the maximum potential, and 0 indicates both overtraining and lack of form.

5 Results and Discussion

The final ranking of A_1–A_3 competitors is presented in Fig. 3. Taking into account the determined factors in the model, the most significant increase in the form to the starting point was achieved by the A_1 competitor. Comparing the other two athletes, the A_3 swimmer achieved a smaller form increase, but this is due to a higher entry threshold. This means that at the beginning of the preparations A_3 swimmer was already in a better disposition than swimmer A_2. Professional athletes at a high level will have less progress from season to season because with such significant results, it is much harder to improve their best performance.

On the other hand, Fig. 4 shows the results obtained for the decision matrix with randomized data. From the preference values obtained, it can be seen that the A_4 and A_8 athletes were overtired in the final period of preparation for the competition. It is shown by the model's output coefficients of 0.00. It is confirmed by the fact that

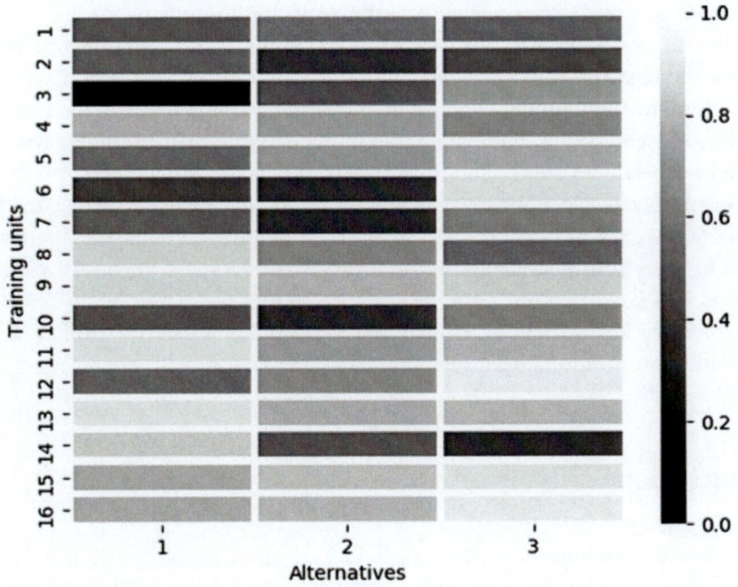

Fig. 3 Model assessments preference values for professional swimmers

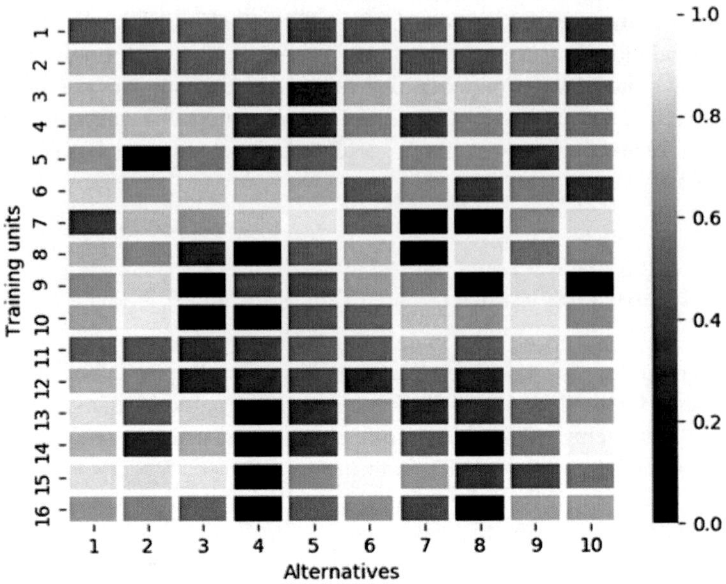

Fig. 4 Model assessments preference values for randomize alternatives A_1–A_{10}

their heart rates at this stage were significantly higher than those of the others, and their decrease did not differ significantly from one measurement to the next. The athlete A_{10} recorded the most significant progress among the analysed athletes. His initial period of preparation was at a sufficiently high intensity. As the number of days to the target competition decreased, the training load reduction was sufficient to achieve the best starting prognosis among the swimmers analysed.

An optimal training process should include high training loads in the initial phase of preparation, which should translate into high heart rate measurements with a small decrease in subsequent steps. The final preparation stage is characterized by a rapid reduction of successive numerical values in heart rate measurements. In the case of the A_4 and A_8 swimmers considered for the randomized data, they performed at the limit of their abilities during the final preparation stages. If, on the other hand, this period had been worked with attention to more remarkable recovery, it would have been possible for them to achieve better results compared to athlete A_{10}, who obtained the best preference score.

6 Conclusions

The problem of reaching peak form in the target sport event period is a significant topic that many coaches are looking at. They are responsible for the performance of their team members, they are responsible for the preparation stage, and the results achieved. One of the ways to control the progress of a competitor's starting form is to monitor the heart rate in each preparation period. For this problem, this article proposes a model of the assessment of the professional swimmers form based on heart rate analysis. The MCDA method called COMET and linear regression method were used to solve the problem. Moreover, this attempt proved to be effective.

The theory of fuzzy numbers combined with the COMET method was used to create an evaluation model with complete knowledge and a certain degree of uncertainty. As a result of the study, a practical model was developed to assess the heart rate ratio. We consider a heartbeat measures after the task to the number of days remaining to the main start. In the presented model, there were 81 characteristic objects, which required 3240 pairwise comparisons. Three competitors preparing for the national competition in 2018 were examined, and then 10 alternatives for these competitors were subjected to the same test. Among the smaller group, the results showed an increasing tendency. In contrast, the results showed that some of the competitors had reached an optimal starting form among the other group, and some of them were overtrained, which did not allow them to achieve the planned result. It is worth mentioning that the proposed approach can be used in practice by swim coaches as support to visualize the directional tendency of the swimmer's form.

For further research, be aware that only exemplary usage was presented in this paper. A literature-based query and identification of all criteria influencing the form of swimmers should be made. Another challenge is data collection and obtaining a sufficiently large observation base, which can result in the development of a useful expert system supporting the entire training cycle. Besides the heart rate analysis and the number of days to competition, it is also possible to take into account the intensity of training, followed by heart rate measurement, and the achieved level of acidification of the body, showing the degree of realization of the training. In the methodical dimension, the challenge is to remove/reduce the inconvenience of comparing a large number of objects characteristic of the COMET method, which becomes particularly important with the increase in the dimensionality of the model being developed.

References

1. Anderson, M.E., Hopkins, W.G., Roberts, A.D., Pyne, D.B.: Monitoring seasonal and long-term changes in test performance in elite swimmers. Eur. J. Sport. Sci. **6**(3), 145–154 (2006)
2. Aubert, A.E., Seps, B., Beckers, F.: Heart rate variability in athletes. Sport. Med. **33**(12), 889–919 (2003)
3. Blanco, V., Salmerón, R., Gómez-Haro, S.: A multicriteria selection system based on player performance: case study–the Spanish ACB basketball league. Group Decis. Negot. **27**(6), 1029–1046 (2018)
4. Boender, C.G.E., De Graan, J.G., Lootsma, F.A.: Multi-criteria decision analysis with fuzzy pairwise comparisons. Fuzzy Sets Syst. **29**(2), 133–143 (1989)
5. Brans, J.-P., Vincke, P.: Note–a preference ranking organisation method: (the PROMETHEE method for multiple criteria decision-making). Manag. Sci. **31**(6), 647–656 (1985)

6. Cinelli, M., Kadziński, M., Gonzalez, M., Słowiński, R.: How to support the application of multiple criteria decision analysis? Let us start with a comprehensive taxonomy. Omega 102261 (2020)
7. Dadelo, S., Turskis, Z., Zavadskas, E.K., Dadeliene, R.: Multi-criteria assessment and ranking system of sport team formation based on objective-measured values of criteria set. Expert Syst. Appl. **41**(14), 6106–6113 (2014)
8. Daglioglu, O.: The effect of gradually increasing exercise on oxygen consumption and lactate levels in swimmers. Ann. Biol. Res. **4**(10), 96–102 (2013)
9. Deschrijver, G., Kerre, E.E.: On the relationship between some extensions of fuzzy set theory. Fuzzy Sets Syst. **133**(2), 227–235 (2003)
10. Faizi, S., Rashid, T., Sałabun, W., Zafar, S., Wątróbski, J.: Decision making with uncertainty using hesitant fuzzy sets. Int. J. Fuzzy Syst. **20**(1), 93–103 (2018)
11. Faizi, S., Sałabun, W., Ullah, S., Rashid, T., Więckowski, J.: A new method to support decision-making in an uncertain environment based on normalized interval-valued triangular fuzzy numbers and comet technique. Symmetry **12**(4), 516 (2020)
12. Faude, O., Meyer, T., Scharhag, J., Weins, F., Urhausen, A., Kindermann, W.: Volume vs. intensity in the training of competitive swimmers. Int. J. Sport. Med. **29**(11), 906–912 (2008)
13. Grant, S., McMillan, K., Newell, J., Wood, L., Keatley, S., Simpson, D., Leslie, K., Fairlie-Clark, S.: Reproducibility of the blood lactate threshold, 4 mmol· 1-1 marker, heart rate and ratings of perceived exertion during incremental treadmill exercise in humans. Eur. J. Appl. Physiol. **87**(2), 159–166 (2002)
14. Guitouni, A., Martel, J.-M.: Tentative guidelines to help choosing an appropriate MCDA method. Eur. J. Oper. Res. **109**(2), 501–521 (1998)
15. Gupta, M.M., Qi, J.: Theory of t-norms and fuzzy inference methods. Fuzzy Sets Syst. **40**(3), 431–450 (1991)
16. Hashem, M.M.A., Shams, R., Kader, M.A., Sayed, M.A.: Design and development of a heart rate measuring device using fingertip. In: International Conference on Computer and Communication Engineering (ICCCE'10), pp. 1–5. IEEE (2010)
17. Heck, H., Mader, A., Hess, G., Mücke, S., Müller, R., Hollmann, W.: Justification of the 4-mmol/l lactate threshold. Int. J. Sport. Med. **6**(03), 117–130 (1985)
18. Hedelin, R., Kenttá, G., Wiklund, U., Bjerle, P., Henriksson-Larsén, K.: Short-term overtraining: effects on performance, circulatory responses, and heart rate variability. Med. Sci. Sport. Exerc. **32**(8), 1480–1484 (2000)
19. Hellard, P., Guimaraes, F., Avalos, M., Houel, N., Hausswirth, C., Toussaint, J.F.: Modeling the association between HR variability and illness in elite swimmers. Med. Sci. Sport. Exerc. **43**(6), 1063 (2011)
20. Hooper, S.L., Mackinnon, L.T.: Monitoring regeneration in elite swimmers. In: Overload, Performance Incompetence, and Regeneration in Sport, pp. 139–148. Springer, Berlin (1999)
21. Kellmann, M.: Preventing overtraining in athletes in high-intensity sports and stress/recovery monitoring. Scand. J. Med. Sci. Sport. **20**, 95–102 (2010)
22. Kiviniemi, A.M., Hautala, A.J., Kinnunen, H., Tulppo, M.P.: Endurance training guided individually by daily heart rate variability measurements. Eur. J. Appl. Physiol. **101**(6), 743–751 (2007)
23. Kizielewicz, B., Kołodziejczyk, J.: Effects of the selection of characteristic values on the accuracy of results in the comet method. Procedia Comput. Sci. **176**, 3581–3590 (2020)
24. Kizielewicz, B., Sałabun, W.: A new approach to identifying a multi-criteria decision model based on stochastic optimization techniques. Symmetry **12**(9), 1551 (2020)
25. Kizielewicz, B., Szyjewski, Z.: Handling economic perspective in multicriteria model-renewable energy resources case study. Procedia Comput. Sci. **176**, 3555–3562 (2020)
26. Kizielewicz, B., Dobryakova, L.: How to choose the optimal single-track vehicle to move in the city? Electric scooters study case. Procedia Comput. Sci. **176**, 2243–2253 (2020)
27. Kizielewicz, B., Dobryakova, L.: MCDA based approach to sports players' evaluation under incomplete knowledge. Procedia Comput. Sci. **176**, 3524–3535 (2020)

28. Kubler, S., Robert, J., Derigent, W., Voisin, A., Le Traon, Y.: A state-of the-art survey & testbed of fuzzy AHP (FAHP) applications. Expert Syst. Appl. **65**, 398–422 (2016)
29. Leclercq, J.: Propositions d'extension de la notion de dominance en présence de relations d'ordre sur les pseudo-critéres: la méthode melchior. JORBEL-Belg. J. Oper. Res. Stat. Comput. Sci. **24**(1), 32–46 (1984)
30. Londeree, B.R.: Effect of training on lactate/ventilatory thresholds: a meta-analysis. Med. Sci. Sport. Exerc. **29**(6), 837–843 (1997)
31. Mourot, L., Bouhaddi, M., Perrey, S., Cappelle, S., Henriet, M.-T., Wolf, J.-P., Rouillon, J.-D., Regnard, J.: Decrease in heart rate variability with overtraining: assessment by the poincare plot analysis. Clin. Physiol. Funct. Imaging **24**(1), 10–18 (2004)
32. Mujika, I.: Quantification of training and competition loads in endurance sports: methods and applications. Int. J. Sport. Physiol. Perform. **12**(s2), S2-9 (2017)
33. Pastijn, H., Leysen, J.: Constructing an outranking relation with Oreste. In: Models and Methods in Multiple Criteria Decision Making, pp. 1255–1268. Elsevier, Amsterdam (1989)
34. Piegat, A.: Fuzzy Modeling and Control, vol. 742. Physica (2001)
35. Piegat, A., Sałabun, W.: Nonlinearity of human multi-criteria in decision-making. J. Theor. Appl. Comput. Sci. **6**(3), 36–49 (2012)
36. Psycharakis, S.G.: A longitudinal analysis on the validity and reliability of ratings of perceived exertion for elite swimmers. J. Strength Cond. Res. **25**(2), 420–426 (2011)
37. Riaz, M., Sałabun, W., Farid, H.M.A., Ali, N., Wątróbski, J.: A robust q-rung orthopair fuzzy information aggregation using Einstein operations with application to sustainable energy planning decision management. Energies **13**(9), 2155 (2020)
38. Roubens, M.: Fuzzy sets and decision analysis. Fuzzy Sets Syst. **90**(2), 199–206 (1997)
39. Roy, B.: Multicriteria Methodology for Decision Aiding, vol. 12. Springer Science & Business Media, Berlin (2013)
40. Sałabun, W.: The mean error estimation of TOPSIS method using a fuzzy reference models. J. Theor. Appl. Comput. Sci. **7**(3), 40–50 (2013)
41. Sałabun, W.: The characteristic objects method: a new distance-based approach to multicriteria decision-making problems. J. Multi-Criteria Decis. Anal. **22**(1–2), 37–50 (2015)
42. Sałabun, W., Karczmarczyk, A.: Using the comet method in the sustainable city transport problem: an empirical study of the electric powered cars. Procedia Comput. Sci. **126**, 2248–2260 (2018)
43. Sałabun, W., Karczmarczyk, A., Wątróbski, J., Jankowski, J.: Handling data uncertainty in decision making with comet. In: 2018 IEEE Symposium Series on Computational Intelligence (SSCI), pp. 1478–1484. IEEE (2018)
44. Sałabun, W., Piegat, A.: Comparative analysis of MCDM methods for the assessment of mortality in patients with acute coronary syndrome. Artif. Intell. Rev. **48**(4), 557–571 (2017)
45. Sałabun, W., Wątróbski, J., Shekhovtsov, A.: Are MCDA methods benchmarkable? A comparative study of TOPSIS, VIKOR, COPRAS, and PROMETHEE II methods. Symmetry **12**(9), 1549 (2020)
46. Sałabun, W., Ziemba, P., Wątróbski, J.: The rank reversals paradox in management decisions: the comparison of the AHP and comet methods. In: International Conference on Intelligent Decision Technologies, pp. 181–191. Springer (2016)
47. Seiler, S.: What is best practice for training intensity and duration distribution in endurance athletes? Int. J. Sport. Physiol. Perform. **5**(3), 276–291 (2010)
48. Shekhovtsov, A., Kozlov, V., Nosov, V., Sałabun, W.: Efficiency of methods for determining the relevance of criteria in sustainable transport problems: a comparative case study. Sustainability **12**(19), 7915 (2020)
49. Smith, D.J., Norris, S.R., Hogg, J.M.: Performance evaluation of swimmers. Sport. Med. **32**(9), 539–554 (2002)
50. Streiner, D.L., Norman, G.R., Cairney, J.: Health measurement scales: a practical guide to their development and use. Oxford University Press, Oxford (2015)
51. Tiwari, R.K., Kumar, R.: G-TOPSIS: a cloud service selection framework using Gaussian TOPSIS for rank reversal problem. J. Supercomput. (2020)

52. Wątróbski, J., Jankowski, J., Ziemba, P., Karczmarczyk, A., Zioło, M.: Generalised framework for multi-criteria method selection. Omega **86**, 107–124 (2019)
53. Watróbski, J., Sałabun, W.: The characteristic objects method: a new intelligent decision support tool for sustainable manufacturing. In: International Conference on Sustainable Design and Manufacturing, pp. 349–359. Springer (2016)
54. Więckowski, J., Kizielewicz, B., Kołodziejczyk, J.: Application of hill climbing algorithm in determining the characteristic objects preferences based on the reference set of alternatives. In: International Conference on Intelligent Decision Technologies, pp. 341–351. Springer (2020)
55. Więckowski, J., Kizielewicz, B., Kołodziejczyk, J.: Finding an approximate global optimum of characteristic objects preferences by using simulated annealing. In: International Conference on Intelligent Decision Technologies, pp. 365–375. Springer (2020)
56. Więckowski, J., Kizielewicz, B., Kołodziejczyk, J.: The search of the optimal preference values of the characteristic objects by using particle swarm optimization in the uncertain environment. In: International Conference on Intelligent Decision Technologies, pp. 353–363. Springer (2020)
57. Zadeh, L.A.: Fuzzy sets. Inf. Control **8**(3), 338–353 (1965)
58. Zhü, K.: Fuzzy analytic hierarchy process: fallacy of the popular methods. Eur. J. Oper. Res. **236**(1), 209–217 (2014)
59. Zimmermann, H.-J.: Fuzzy Set Theory—and Its Applications. Springer Science & Business Media, Berlin (2011)

Brown–Gibson Model as a Multi-criteria Decision Analysis (MCDA) Method: Theoretical and Mathematical Formulations, Literature Review, and Applications

Nasser Yimen, Theodore Tchotang, Abraham Kanmogne, Yungho Adamu, Fombe Lawrence Fon, and Mustafa Dagbasi

Abstract Over the past four decades, multicriteria decision analysis (MCDA) has become an essential discipline for operations research. It has been beneficial to address the insufficiency of a single criterion in real-world decision-making. MCDA models have been classified into three main categories: Multi-Attribute Decision Analysis (MADA), Multi-Objective Decision Analysis (MODA), and the combination of the two. Among the MADA models, the Brown–Gibson model is gaining more and more academic attention. In 1972, Phillip A. Brown and David F. Gibson designed it to consider both objective and subjective factors in optimal location problems. Since then, it has been applied in various engineering and science fields, and different versions of the model have been established. However, no literature review study related to this model has been performed to date. Such an analysis will provide researchers wishing to embark on related studies with a solid background. The present book chapter aims to fill this gap by presenting the theoretical and mathematical formulations of different versions of the model listed from the literature and carrying out a literature review. The latter showed that most of the applications took place in Asia, wherein either the original or extended version of the model was used. By way of illustration, an application of the original version to determine the optimal location of a commercial centre in Cameroon showed that the city of Douala was the best location.

Keywords Operations research · Multicriteria decision analysis (MCDA) · Multi-Attribute Decision Analysis (MADA) · Brown–Gibson model · Optimal location · Subjective factors · Objective factors

---

N. Yimen (✉) · T. Tchotang · A. Kanmogne
National Advanced School of Engineering, University of Yaoundé I, Yaoundé, Cameroon

N. Yimen · Y. Adamu · F. L. Fon
The Higher Institute of Transport and Logistics (HITL), University of Bamenda, Bamenda, Cameroon

M. Dagbasi
Department of Energy Systems Engineering, Cyprus International University, Nicosia, Cyprus
e-mail: mdagbasi@ciu.edu.tr

A. J. Kulkarni (ed.), *Multiple Criteria Decision Making*, Studies in Systems, Decision and Control 407, https://doi.org/10.1007/978-981-16-7414-3_9

Abbreviation and Mathematical Symbols

AHP	Analytical Hierarchy Process
AMT	Advanced Manufacturing Technology
APM	Assessment Preference Measure
AS/RS	Automatic Storage and Retrieval System
CFI	Critical Factor Index
CFM	Critical Factor Measure
CSM	Computer Software Measure
CT	Conventional Technology
CTE	Cost and Time Effectiveness
DSS	Decision Support System
EBG	Extended Brown–Gibson
EC	Effective Cost
ENF	Effective Non-Financial
ET	Effective Time
FLSI	Fuzzy Location Selection Index
FOFC	Fuzzy Objective Factor of Alternative
FOFM	Fuzzy Objective Factor Measure
FSFM	Fuzzy Subjective Factor Measure
IEC	Ineffective Cost
IET	Ineffective Time
INF	Ineffective Non-Financial
LM	Location Measure
LPM	Location Preference Measure
MADA	Multi-Attribute Decision Analysis
MCDA	Multi-Criteria Decision Analysis
MODA	Multi-Objective Decision Analysis
MSPM	Manufacturing System Preference Measure
NPV	Net Present Value
OFC	Objective Factor Cost
OFM	Objective Factor Measure
PSPM	Pumping System Preference Measure
QFD	Quality Function Deployment
RMS	Reconfigurable Manufacturing Systems
SFM	Subjective Factor Measure
TS	Traditional System
TVM	Time Value of Money
W	The subjective factor weight
α	The objective factor decision weight

1 Introduction

Multi-criteria decision analysis (MCDA) has become a significant discipline in the operations research area over the past four decades. The rise of MCDA in decision-making processes is based on the simple conclusion that a single goal, objective, criterion or point of view is often insufficient to make decisions in the real world [34]. The field of MCDA consists of developing appropriate approaches to support and assist decision-makers in situations where they have to take into account several conflicting decision-making factors (objectives, criteria, goals, etc.) simultaneously [35].

Typically, MCDA problems consist of five elements: objective, decision-makers' preferences, alternatives, criteria, and outcomes [19]. MCDA models can be classified into three main categories: Multi-Attribute Decision Analysis (MADA), Multi-Objective Decision Analysis (MODA), and the combination of both, as shown in Fig. 1 [22].

The difference between MADA and MODA can be established based on the number of alternatives considered. They are continuous (infinite) in MODA and discrete (finite) in MADA. Otherwise, both share the same functionality [13].

Among the MADA models, the Brown–Gibson model is gaining more and more attention from scholars. It was initially designed in 1972 by Phillip A. Brown and David F. Gibson for optimal plant location problems [5]. At that time, the existing plant location models presented the limitation of not combining quantitative and qualitative factors that might affect the suitability of an alternative location for a particular plant. Two classes of methods existed in the literature, namely, quantitative models and qualitative models. Quantitative methods were mostly cost or profit-oriented and included the following four concepts: (i) least total cost location, (ii) least production cost location, (iii) least distribution cost location, and (iv) maximum profit location.

The most applied qualitative model at that time was the location scoring model. This model consists of four steps [16]:

1. Identifying all factors that can affect the suitability of alternative sites.
2. Assigning weights to each factor based on its relative importance.
3. Assigning a score to each location for each factor.

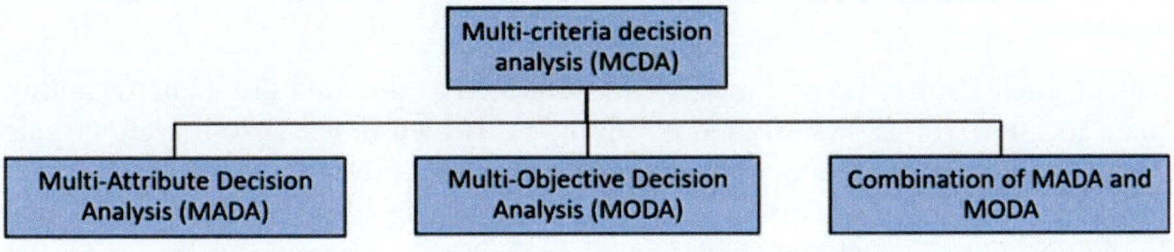

Fig. 1 Classification of multi-criteria decision analysis (MCDA) models [22]

4. Computing the index of each candidate location by performing the weighted sum on all identified factor's scores and selecting the location with the highest index.

Nowadays, this method remains very popular.

The main disadvantage of the qualitative model presented above is that it is based entirely on a subjective evaluation to make a location decision. Although this drawback is overcome by the quantitative models listed above, the latter does not incorporate non-quantifiable factors that can significantly impact the plant location decision.

The above lines draw the context in which the Brown–Gibson model was proposed as a single-site and multi-attribute model to address the disadvantages of qualitative and quantitative methods while combining their strengths.

Although the model was initially dedicated to location decision problems, it has been applied in various engineering and science areas such as manufacturing system selection, supplier selection, service performance measurement, or software package selection. Besides, different versions of the model have been established to address the original model's limitations or adapt the model to a particular application situation. To the best of the authors' knowledge, no literature review study on the Brown–Gibson model has been performed in the literature until date. This book chapter aims to fill the gap by carrying out a literature review on the Brown–Gibson model while presenting the theoretical and mathematical formulations of different versions of the model listed in the literature.

The rest of this book chapter is organized as follows: Sect. 2 outlines the original Brown–Gibson model's theoretical and mathematical formulations and its different modified versions listed from the literature. Section 3 presents a systematic literature review of studies based on the Brown–Gibson model highlighting the purpose, application area and version of the model used in each study. Then, an application of the model's original version is outlined for illustration purposes in Sect. 4. The book chapter ends with a conclusion in Sect. 5.

2 Theoretical and Mathematical Formulations

2.1 Original Brown–Gibson Model

The original Brown–Gibson model was dedicated to decision problems regarding plant location. It was first drafted by Phillip A. Brown in his 1970 master's thesis entitled "Plant location: a quantified model for community and plant site selection" at Montana State University, USA [6]. The model was then formally established by Phillip A. Brown and David F. Gibson in their 1972 scientific article, under the title "A Quantified Model for Facility Site Selection—Application to a Multiplant Location Problem" [5].

Four stages underpinned the development of this model: (i) classification of location factors, (ii) definition of the general model in terms of all types of factors, (iii) quantification of general terms of the model, and (iv) formulation of the final model.

2.1.1 Classification of Location Factors

Brown and Gibson [5] classified factors affecting the suitability of locations for plant siting into three categories, namely, (i) critical factors, (ii) objective factors and (iii) subjective factors. Notwithstanding other favourable conditions, critical factors prevent installing a plant on a site if they are not observed. For example, a water-intensive industry such as a brewery will not establish itself at a location with potential water shortages, regardless of low labour or raw material costs. Objective factors are those that can be assessed on a monetary basis, such as the cost of raw materials, labour or transport. Finally, subjective factors are characterized by a qualitative type measure and can rarely be assigned a monetary value. For example, the attitude of the community towards setting up a factory is a subjective factor. It should be noted that a factor can be categorized as both critical and objective or critical and subjective.

2.1.2 Definition of the General Model

A combination of the three factors described above is the basis for the general model. For each alternative location *i*, a location measure, LM_i, is specified as follows:

$$LM_i = CFM_i[\alpha \times OFM_i + (1 - \alpha) \times SFM_i] \tag{1}$$

where CFM_i ($CFM_i = 0$ *or* 1), OFM_i ($0 \leq OFM_i \leq 1$, and $\sum_i OFM_i=1$) and SFM_i ($0 \leq SFM_i \leq 1$, and $\sum_i SFM_i=1$) denote the critical factor measure, objective factor measure and subjective factor measure, respectively. α ($0 \leq \alpha \leq 1$) is the objective factor decision weight. These terms are discussed below.

Alternative sites for plant installation are listed in descending order of location measure—the one with the highest location measure being the best.

2.1.3 Quantification of Location Measures

The critical factor measure of the alternative site *i*, CFM_i is expressed as follows:

$$CFM_i = \prod_j CFI_{ij}, i = 1, 2, \ldots, n, \tag{2}$$

where CFI_{ij}, the critical factor index for location *i* with respect to the critical factor CF_j, takes the value 1 if location *i* meets the requirement of the critical factor CF_j, and 0 otherwise. Thus, if any critical factor index is zero, CFM_i and LM_i will also be zero, and site i will therefore be excluded from the ranking.

The objective factor measure and the subjective factor measure are dimensionless indices that help ensure consistency between subjective and objective factors.

For each alternative site i, the objective factor measure OFM_i is defined as:

$$OFM_i = \left[OFC_i \times \sum_i (1/OFC_i)\right]^{-1} \quad (3)$$

where OFC_i denotes the total objective factor cost of the alternative site i. It represents the sum of all objective factors in monetary terms ($\sum_j OF_{ij}$) related to setting the plant at site i.

The subjective factor measure (SFM) for each alternative site i is defined in Eq. (4):

$$SFM_i = \sum_k (W_k \times SF_{ik}) \quad (4)$$

where the subjective factor weight (W_j) and the site ranking for the subjective factor SF_k (SF_{ik}) values are systematically assigned with the help of preference theory or forced-choice pairwise, a subjective quantification technique. The latter is based on a pairwise comparison of factors and on the construction of a general preference matrix. By comparing two factors, three scenarios are possible: (i) the first factor is more important than the second (the corresponding numerical values are 1 for the first factor and 0 for the second), (ii) the second factor is more important than the first (the corresponding numerical values are 1 for the second factor and 0 for the first), (iii) both factors have the same importance (both factors receive 1). Further detailed information on preference theory with illustration is available in [5] and [11].

The last term α $(0 \le \alpha \le 1)$ in Eq. (1) is the objective factor decision weight. It is determined by the management and represents the relative importance of the objective factors compared with subjective factors of the location problem.

Given the high degree of subjectivity involved in determining this relative weight, the sensitivity analysis of location measure to the variation of this parameter is most often recommended.

2.1.4 Formulation of the Final Model

Based on the developments in previous sub-sub-sections, the final model defining the location measure LM_i for each site i can be expressed as follows:

$$LM_i = \left\{\prod_j CFI_{ij}\right\}$$

$$\times\left\{\alpha\times\left[OFC_i\times\sum_i(1/OFC_i)\right]^{-1}+(1-\alpha)\times\sum_k(W_k\times SF_{ik})\right\} \tag{5}$$

The optimal site is the one that receives the highest location measure as expressed by Eq. (5)

2.2 *Buffa and Sarin Version of Brown–Gibson Model*

To deal with the complexity of the original Brown–Gibson model, [7] proposed a simpler version of the model in their textbook entitled "Modern Production/Operations Management", edited by John Willey & Sons. Inc. in 1987. In this model, for each alternative site *i*, the objective factor measure is obtained by converting the total objective factor cost $\sum_j OF_{ij}$ into a dimensionless form by using the normalization formula as follows:

$$OFM_i=\frac{max_i\left[\sum_j OF_{ij}\right]-\sum_j OF_{ij}}{max_i\left[\sum_j OF_{ij}\right]-min_i\left[\sum_j OF_{ij}\right]} \tag{6}$$

In this model, the site with the maximum total objective factor cost receives zero as the objective factor measure value. In contrast, the one with the minimum total objective factor receives one as the objective factor measure value. Besides, for each alternative site *i*, the numerical values of each subjective factor *k*, SF_{kj} is assigned on a 0–1 scale based on the judgement of the decision-maker. The relative weight of subjective factors, w_k is also assigned based on the decision-maker judgement. The final model is then expressed as follows:

$$LM_i=\prod_j CFI_{ij}$$
$$\times\left[\alpha\times\frac{max_i\left[\sum_j OF_{ij}\right]-\sum_j OF_{ij}}{max_i\left[\sum_j OF_{ij}\right]-min_i\left[\sum_j OF_{ij}\right]}+(1-\alpha)\times\sum_k(SF_{ki}\times W_k)\right] \tag{7}$$

While this version of Brown–Gibson model is more straightforward, consistency in assigning values and weights to subjective factors is not guaranteed.

2.3 The Extended Brown–Gibson Model

One of the shortcomings of the previous versions of the Brown–Gibson model is that objective factors only incorporate cost-related aspects and do not spell out how to integrate other quantitative factors that may impact an alternative's suitability. Such elements may include time factors (to be minimized or maximized), profit (to be maximized) as well as other non-financial tangible factors (to be minimized or maximized). To fill this gap, [1] presented the Extended Brown–Gibson model, which categorizes objective factors into effective factors (to be maximized) and ineffective factors (to be minimized). They defined the cost and time effectiveness of alternative *i*, *CTEi*, as follows:

$$CTE_i = EC_i \frac{1}{\sum_i EC_i} + \left[IEC_i \sum_i \frac{1}{IEC_i} \right]^{-1} + ET_i \frac{1}{\sum_i ET_i} + \left[IET_i \sum_i \frac{1}{IET_i} \right]^{-1} \quad (8)$$

where:

EC_i	Effective cost for alternative *i*.
IEC_i	Ineffective cost for alternative *i*.
ET_i	Effective time for alternative *i*.
IET_i	Ineffective time for alternative *i*.

From *CTEi*, the objective factor measure of each alternative *i*, OFM_i was expressed as follows:

$$OFM_i = CTE_i \frac{1}{\sum_i CTE_i} \quad (9)$$

Finally, they expressed the manufacturing system preference measure of alternative *i*, $MSPM_i$ as:

$$MSPM_i = \prod_j CFI_{ij} \times \left[\alpha \times \left(CTE_i \frac{1}{\sum_i CTE_i} \right) + (1-\alpha) \times \sum_k (SF_{ki} \times W_k) \right] \quad (10)$$

Later, [27] defined a more extended version by considering other non-financial tangible factors and defined the cost and time effectiveness of alternative *i*, *CTEi* as follows:

$$CTE_i = EC_i \frac{1}{\sum_i EC_i} + \left[IEC_i \sum_i \frac{1}{IEC_i} \right]^{-1} + ET_i \frac{1}{\sum_i ET_i} + \left[IET_i \sum_i \frac{1}{IET_i} \right]^{-1}$$

$$+ ENF_i \frac{1}{\sum_i ENF_i} + \left[INF_i \sum_i \frac{1}{INF_i} \right]^{-1} \tag{11}$$

where:

ENF_i Effective non-financial for alternative i.
INF_i Ineffective non-financial for alternative i.

2.4 Yimen and Dagbasi Version of Brown–Gibson Model

The previous versions of the Brown–Gibson model are disadvantaged because they do not consider the time value of money (TVM), making their application limited. Indeed, for most organizations and systems, the resulting cash flows are distributed over a long period [37, 38]. Yimen and Dagbasi [36] took this reality into account by discounting all objective factors (cash flows) associated with each alternative i, resulting in the substitution of the sum of objective factors by the opposite of the net present value (NPV) in Buffa and Sarin version of Brown–Gibson model:

$$LM_i = \prod_j CFI_{ij} \times \left[\alpha \times \frac{NPV_i - min_i[NPV_i]}{max_i[NPV_i] - min_i[NPV_i]} + (1-\alpha) \times \sum_k (SF_{ki} \times W_k) \right] \tag{12}$$

2.5 Analytical Hierarchy Process (AHP)-Integrated Brown Gibson Model

Another shortcoming of the original Brown–Gibson model is related to the characteristics of the preference theory used to evaluate subjective factors. Indeed, the preference theory is not flexible enough when comparing two factors. It allows only two options: (i) one factor is more important than another or (ii) both factors have the same importance. Besides, it is not possible to check the consistency of the judgements of the decision-maker. To overcome these drawbacks, most recent Brown–Gibson model applications used the Analytical Hierarchy Process (AHP) to assess subjective factors instead of the preference theory or forced-choice pairwise comparison procedure.

The Analytic Hierarchy Process (AHP) is an approach developed in the 1970s by Thomas Saaty. Pairwise comparisons are performed at each stage of the criteria' hierarchical structure, leading to the building decision matrices. The values used for pairwise comparison are integer values between 1 and 9 or their reciprocal values. Detailed information related to the AHP method is available in [31].

2.6 *Fuzzy Brown–Gibson Model*

The fuzzy Brown–Gibson model was designed by [4]. It is based on fuzzy triangular numbers to provide more practicality, make the site selection problem more realistic and facilitate managerial understanding. Detailed information about fuzzy triangular numbers is available in [10] and [20]. In this model, the objective function, the fuzzy location selection index (FLSI) for each alternative site is defined in Eq. (13):

$$FLSI_i = CFM_i[\alpha \times FOFM_i + (1 - \alpha) \times FSFM_i] \tag{13}$$

where $FOFM_i$ and $FSFM_i$ are, respectively, the fuzzy objective factor measure and fuzzy subjective factor measure of the alternative site *i*. $FOFM_i$ is defined as follows:

$$FOFM_i = \left[FOFC_i \times \sum\nolimits_j (1/FOFC_j)\right]^{-1} \tag{14}$$

where $FOFC_i$ = Fuzzy objective factor of alternative *i*.

3 Litterature Review

Numerous studies in different countries and sectors applying Brown–Gibson Model have been performed in the literature. Most of these applications took place in Asia and used the original or extended version of the model. Among those who used the model's original version, [32] applied this version to select the best financial accounting software package. They defined the Computer Software Measure (CSM) as the objective function in terms of two critical factors (Hardware compatibility and Operational compatibility), four objective factors (information quality, provision of audit trails, "track record" of vendor, expandability, and quality of security system) and three subjective factors (initial purchase cost, costs of training users and operators, cost of operation, maintenance and modification). Hemalatha et al. [14] applied the same version to find the best location for a retail store among four alternative locations in India, namely, Thiruvarembur, Cantonment, Thillainagar and Srirangam. They defined the Location Preference Measure (LPM) as an objective function based on four subjective factors (population, availability of skilled labour, low-risk environment and availability of store location) and two objective factors (total rent per year and the total labour cost per year). Dominic et al. [9] used the original Brown–Gibson model to select the best location for offshore outsourcing of

Malaysia's typical software company. They defined the location measure preference (LMP) as the objective function in their analysis and considered five alternative locations: India, USA, China, Malaysia and the Philippines. Using the same objective function, [15] developed a suitable decision support system (DSS) for retail location decision. They integrated into the designed DSS five subjective factors (population, availability of skilled labour, competition, economic base and legal characteristics) and two objective factors (rent per month and labour cost per month). Rahman et *al.* [29] also developed a similar DSS to determine the optimal location for an automobile manufacturing company in Bangladesh considering four alternative sites: Chittagong, Mongla, Kustia, Narayongonj and Rangpur. All the selected research studies that applied the original version of Brown–Gibson model are summarized in Table 1.

Compared with the original model, the extended Brown–Gibson (EBG) model was more applied in the literature. Most of the EBG model-related studies are dedicated to the manufacturing sector like the one by [1]. They applied the model to justify an investment in advanced manufacturing technology (AMT), a new technology for enhancing manufacturing systems' efficiency and flexibility. They defined the manufacturing system preference measure (MSPM) as the objective function for comparing the new AMT and the traditional system (TS). Ragavan and Punniyamoorthy [28] conducted a similar analysis except that they used the AHP to assess subjective factors instead of preference theory. Later, [27] followed the same methodology to justify the use of the automatic storage and retrieval system (AS/RS) in the heavy engineering industry and [21] did the same to justify an investment in the Reconfigurable Manufacturing Systems (RMS), new technology to enhance manufacturing systems' efficiency and flexibility. Both studies adopted the Manufacturing system preference measure (MSPM) as the objective function and used AHP to evaluate the subjective factors. Bagum et al. [2] and [25] also proposed analytical models based on the extended Brown–Gibson model and Analytical Hierarchy Process to find the best pumping system for irrigation in Bangladesh and the best suppliers of a supply chain network in India, respectively.

In some studies, the authors integrated the Extended Brown–Gibson model with quality function deployment (QFD) to create a closed-loop model to assess and improve system performance. Those studies include the analyses by [23] and [24], who developed closed-loop models for service performance management in automobile repair shops. Parthiban and Goh [26] proposed a closed-loop model to measure and improve a manufacturing management system's performance. The selected studies related to the extended Brown–Gibson model are summarized in Table 2.

Besides studies that applied the original or extended versions of the Brown–Gibson model, [8] used Buffa and Sarin version of the model to determine the best site for constructing a bowling alley in Magusa, Cyprus. Feridun et al. [12] applied the same version of the model for automobile selection. Yimen and Dagbasi [36] combined the version of the model they developed with RETScreen software to determine the optimal location for a 5-MW solar photovoltaic plant in northern Cameroon. Bhattacharya et al. [4], Kaboli et al. (2007), [33] and [18] applied the

Table 1 Original Brown–Gibson model-based selected studies

Authors, countrry	Objective and objective function	Alternatives	Critical factors	Subjective factors	Objective factors
[32], USA	• Selecting the best financial accounting software package • Computer software measure (CSM)	• A, B, C, E, F and G	• Hardware compatibility • Operational compatibility	• Information quality • Provision of audit trails • "Track record" of vendor • Expandability • Quality of security system	• Initial purchase cost • Costs of training users and operators • Costs of operation, maintenance, and modification
[14], India	• Finding the best location for a retail store • Location preference measure (LPM)	• Thiruvarembur • Cantonment • Thillainagar • Srirangam		• Population • Availability of Skilled Labour • Low Risk Environment • Availability of Store Location	• Total rent per year • Total labour cost per year
[9], Malaysia	• Selecting the best location for offshore outsourcing of a typical software company • Location preference Measure (LPM)	• India • USA • China • Malaysia • Philippines		• Quality of labour • Low-risk environment • Language capabilities • Population	• Labour cost • Tax savings
[15], India	• Finding the best location for a retail store • Location preference measure (LPM)	• UAE • China • India • Singapore		• Population • Availability of Skilled Labour • Competition • Economic base • Legal characteristics	• Rent per month • Labour cost per month

(continued)

Table 1 (continued)

Authors, countrry	Objective and objective function	Alternatives	Critical factors	Subjective factors	Objective factors
[29], Bangladesh	• Finding the best location for an an automobile company • Location preference measure (LPM)	• Chittagong • Mongla • Kustia • Narayongonj • Rangpur		• Skill of Worker • Customer proximity • Community attitude • Communication network	• Skill of worker • Customer proximity • Community attitude • Communication network • Other factors

fuzzy Brown–Gibson model to problems of finding the optimal location for facilities in Iran.

4 Application: Original Brown–Gibson Model for a Commercial Centre Location Decision

This section presents an application of the original Brown–Gibson model in determining the optimal location to set a commercial centre in Cameroon. Six alternative locations were identified, namely, Douala, Bafoussam, Yaoundé, Maroua, Kribi and Buea. Besides, we considered in this application the following factors.

1. Critical factors
   - CF_1: Land availability
   - CF_2: Energy availability
   - CF_3: Construction permission
2. Objective factors
   - OF_1: Construction cost
   - OF_2: Land cost
   - OF_3: All other cost
3. Subjective factors
   - SF_1: Population
   - SF_2: Ease of transportation
   - SF_3: Community attitude
   - SF_4: Availability of Skilled Labour

Table 2 Extended Brown–Gibson model based on selected studies

Authors and country	Objective and objective function	Alternatives	Subjective factors	Effective factors	Ineffective factors
[1], India	• Justifying an investment in the AMT technology • Manufacturing system preference measure (MSPM)	• AMT • TS	• Flexibility • Capacity • Learning • Exposure to labour unrest • Increment	• Profit • Labour savings • Inventory Cost • Floor space savings • Materials savings • Utilization Time	• Depreciation • Material cost • Overheads • Labour cost • Cycle time
[28], India	• Justifying an investment in the AMT technology • Manufacturing system preference measure (MSPM)	• AMT • CT	• Flexibility • Capacity • Learning • Exposure to labour unrest • Increment	• Present value of profit	• Present value of annual cost • Present value of production cost • Present value of depreciation
[27], India	• Justifying an investment in AS/RS in a heavy engineering industry • Manufacturing system preference measure (MSPM)	• Existing system of storage • AS/RS	• Avoid handling damages • Safe in operation • Easy identification of materials • Inventory control • Adaptation to FMS • Aesthetic appearance	• Labour saving • Materials saving • Inventory cost saving • Profit • Floor space saving • Utilization time • Productivity	• Labour cost • Material cost • Overheads cost • Cycle time • Service lead time • Machine set-up time • Wastage

(continued)

Table 2 (continued)

Authors and country	Objective and objective function	Alternatives	Subjective factors	Effective factors	Ineffective factors
[30], India	• Selecting the best car • Car preference measure (CPM)	• Chevrolet Aveo • Ford Fiesta • Hyundai Accent • Maruti Baleno	• Seating • Equipment • Ride and Handling • Refinement • Safety and Security	• Performance	• Cost
[21], India	• Justifying an investment in the RMS technology • Manufacturing system preference measure (MSPM)	• RMS • AMS • CMS	• Reconfigurabiltity • Responsiveness • Product cost • Operator skills • Inventory	• Value of profit	• Value of annual cost • Value of production cost • Value of depreciation
[2], Bangladesh	• Finding the best pumping system • Pumping system preference measure (PSPM)	• Wind pump • Electric pump • Diesel pump	• Low risk • Reduced Emission • Operation & Maintenanceability	• Efficiency • Lifetime	• Unit cost of useful energy • Unit cost of water
[25], India	• Finding the best suppliers in a supply chain network • Assessment preference measure (APM)	• Four suppliers (A, B, C and D)	• Supply flexibility • Quickness in delivery • Impression • Reputation	• Profit	• Transportation cost • Delivery (lead) time

(continued)

Table 2 (continued)

Authors and country	Objective and objective function	Alternatives	Subjective factors	Effective factors	Ineffective factors
[23], India	• Measuring the performance of the service performance • Service system performance measure (SSPM)		• Eight service quality factors	• Five effective cost factors • Four effective time factors	• Three ineffective cost factors • Three ineffective time factors
[24], India	• Measuring the performance of the service performance • Service system performance measure (SSPM)		• Six service quality factorsiv	• Five effective cost factors • Four effective time factors	• Three ineffective cost factors • Four ineffective time factors
[26], India	• Measuring the performance of a manufacturing management system • Service system performance measure (SPM)		• Process efficiency • Product quality and customer satisfaction • Product and process innovation	• Cost of goods sold • R&D expenditure • Capacity utilization	• Operating cost per employee • Age of plant and equipment • Rejection ratio

AMT: advanced manufacturing technology; AS/RS: automatic storage and retrieval system; CT: conventional technology; TS: traditional system

The steps below were followed to determine the optimal location of the commercial centre.

Step 1: The critical factor measure (CFM) of each alternative location was calculated. At each location *i*, the value 1 was assigned to the critical factor CF_j if the alternative location met the related requirement of CF_j, and 0 otherwise. For example, the critical factor CF_1 (Land availability) received the value 1 at all alternative, given that the required land for building the commercial centre was available at all alternative locations. On the other hand, the town of Maroua, which was experiencing severe power cuts, received the value 0 for the critical factor CF_2 (Energy availability). The critical factor measure (CFM) for each alternative location was finally computed using Eq. (2). The results are shown in Table 3.

Step 2: The objective factor measure (OFM) of each alternative location for constructing of the commercial centre was calculated at this step. The sum of the three objective factors considered in the analysis was first calculated for each alternative location. Then, the objective factor measure for each location was computed based on Eq. (3). This calculation process of the objective factor measure of each alternative location is summarized in Table 4.

Table 3 Calculation of critical factor measure (CFM) of each location alternative

Alternative location	CF_1	CF_2	CF_3	CFM
Douala	1	1	1	1
Bafoussan	1	1	1	1
Yaoundé	1	1	1	1
Maroua	1	0	1	0
Kribi	1	1	1	1
Buea	1	1	0	0

Table 4 Calculation of the objective factor measure for each alternative location

Items		Alternative location (i)					
		Douala (1)	Bafoussam (2)	Yaoundé (3)	Maroua (4)	Kribi (5)	Buea (6)
OF ($\$10^4$)	OF_1	128	90	125	110	130	115
	OF_2	50	25	45	20	30	26
	OF_3	16	12	18	13	17	14
$OFC_i = \sum OF_i$ ($\$10^4$)		194	127	188	143	177	155
1/OFCj ($\$^{-1} * 10^{-7}$)		5.155	7.874	5.319	6.993	5.65	6.452
$\sum(1/OFC_j)$ ($\$^{-1} * 10^{-7}$)		37.443					
OFM_i		0.138	0.21	0.142	0.187	0.151	0.172

Step 3: At this step, the subjective factor measure for each alternative location is calculated by:

a. Using forced-choice pairwise procedure to compute the relative weight w_j of each subjective factor. One factor is, therefore, chosen over another, or they are rated equal. There were six pairwise comparisons detailed as follows:

    (1) Population versus EAse of transportation: population as more important.
    (2) Population versus community attitude: both are of equal importance.
    (3) Population versus availability of skilled labour: population as more important
    (4) Ease of transportation versus community attitude: both are of equal importance.
    (5) Ease of transportation versus availability of skilled labour: both are of equal importance.
    (6) Community attitude versus availability of skilled labour: community attitude as more important.

    Table 5 summarizes the above pairwise comparisons and the related calculation of the importance weight of each subjective factor w_j.
b. Then, the forced-choice pairwise procedure was also used to determine each site ranking for each subjective factor SF_{ij} ($0 \leq SF_{ij} \leq 1$ and $\sum SF_{ij} = 1$). For that, the pairwise comparison with sites is repeated separately for each subjective factor. These calculations are shown in Tables 6, 7, 8 and 9.
c. Finally, the subjective factor measure of each alternative location i is calculated using Eq. (4). The calculation results are shown in Table 10.

Step 4: At this step, the location preference measure (LPM) of each location is calculated based on Eq. (1) and setting α, the objective factor decision weight is 0.6. The calculation results are shown in Table 11. These final results show that Douala was the optimal location for the commercial centre with an LPM of 0.1872.

Finally, the sensitivity of the location preference measure (LPM) of each alternative location with to the objective factor decision weight was investigated. The results of this sensitivity analysis are shown in Fig. 2.

Table 5 Calculation of relative importance weight of subjective factors

SF_j	Pairwise comparison						Sum of preference	Relative importance
	1	2	3	4	5	6		
Population (SF_1)	1	1	1				3	3/9
Ease of transportation (SF_2)	0			1	1		2	2/9
Community attitude (SF_3)		1		1		1	3	3/9
Availability of Skilled Labour (SF_4)			0		1	0	1	1/9

Table 6 Determination of ranking of each location for population (SF_1)

Site i	Pairwise comparison															Sum of preference	Site ranking
	1	2	3	4	5	6	7	8	9	10	11	12	13	14	15		
Douala	1	1	1	1	1											5	5/21
Bafoussam	0					0	1	0	1							2	2/21
Yaoundé		1				1				1	1	1				5	5/21
Maroua			0				1			0			1	1		3	3/21
Kribi				0				1			0		1		1	3	3/21
Buea					0				1			0		1	1	3	3/21

Table 7 Determination of ranking of each location for EAse of transportation (SF_2)

Site i	Pairwise comparison															Sum of preference	Site ranking
	1	2	3	4	5	6	7	8	9	10	11	12	13	14	15		
Douala	1	1	1	1	1											5	5/17
Bafoussam	0					0	1	0	1							2	2/17
Yaoundé		0				1				1	1	1				4	4/17
Maroua			0				1			0			1	1		3	3/17
Kribi				0				1			0		0		0	1	1/17
Buea					0				0			0		1	1	2	2/17

Table 8 Determination of ranking of each location for community attitude (SF_3)

Site i	Pairwise comparison															Sum of preference	Site ranking
	1	2	3	4	5	6	7	8	9	10	11	12	13	14	15		
Douala	1	1	1	1	1											5	5/20
Bafoussam	0					1	1	0	0							3	3/22
Yaoundé		0				1				1	1	1				4	4/20
Maroua			0				0			1			1	1		3	3/20
Kribi				0				1			0		1		1	3	3/20
Buea					0				1			1		1	0	3	3/20

Table 9 Determination of ranking of each location for availability of skilled labour (SF_4)

Site i	Pairwise comparison															Sum of preference	Site ranking
	1	2	3	4	5	6	7	8	9	10	11	12	13	14	15		
Douala	1	1	1	1	1											5	5/17
Bafoussam	0					1	1	1	1							3	3/17
Yaoundé		0				0				1	1	1				3	3/17
Maroua			0				1			0			1	0		2	2/17
Kribi				0				0			0		1		1	2	2/17
Buea					0				0			0		1	1	2	2/17

Table 10 Summary of subjective factor measure evaluation

Site i	Subjective factor				SFM
	SF_1 (3/9)	SF_2 (2/9)	SF_3 (3/9)	SF_4 (1/9)	
Douala	5/21	5/17	5/20	5/17	0.261
Bafoussam	2/21	2/17	3/20	3/17	0.127
Yaoundé	5/21	4/17	4/20	3/17	0.218
Maroua	3/21	3/17	3/20	2/17	0.15
Kribi	3/21	1/17	3/20	2/17	0.124
Buea	3/21	2/17	3/20	2/17	0.137

Table 11 Location preference measure (LPM) evaluation

	CFM	OFM	SFM	LPM
Douala	1	0.138	0.261	0.1872
Bafoussam	1	0.21	0.127	0.1768
Yaoundé	1	0.142	0.218	0.1724
Maroua	0	0.187	0.15	0
Kribi	1	0.151	0.124	0.1402
Buea	0	0.172	0.137	0

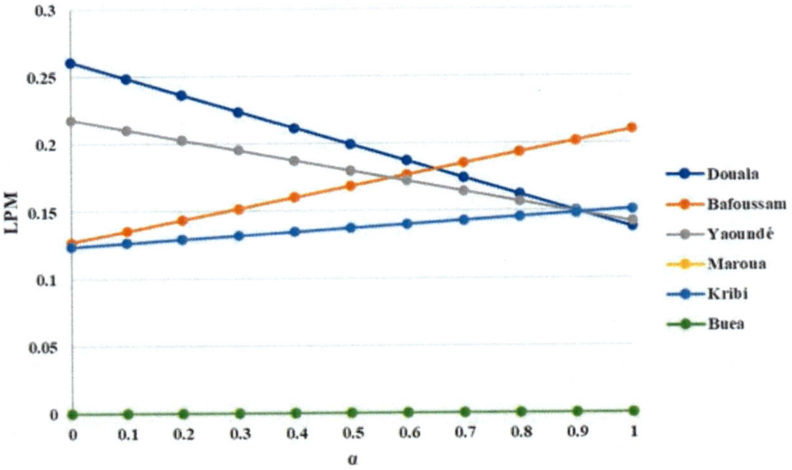

Fig. 2 Sensitivity analysis results of LPM for each location with respect to the objective factor decision weight

Regardless of the objective factor decision weight, the LPMs in Maroua and Buea are zero because their critical factor measures are zero. For values of α less than 6.5, Douala will remain the preferred location, while Bafoussam is the best position for values of α greater than 6.5. These results clearly show the optimality of the Douala location because the appropriate α value is part of the interval [0.3–0.7] [3]. This is in line with the ongoing investment in commercial building construction in the city. In 2020, Central Africa's largest commercial centre, The Douala Grand Mall, was built in the town.

5 Conclusions

This chapter of the book presented different versions of the Brown–Gibson model listed in the literature and reviewed the literature relating to the model. It emerged that the model's original and extended versions were the most used and have been applied to a wide variety of fields. Most of the applications have taken place in Asia, which is not surprising. Globally, Asia is the fastest-growing technology area, with many giant companies like LG, Sony and Samsung.

Applying the original version of the model to determine the best location for a commercial centre in Cameroon highlighted the city of Douala as the most suitable place to establish retail centres.

Two significant facts attest that the Brown–Gibson model is in the process of becoming an essential part of operations research (OR). First, the model has seen several developments and applications since its design. Second, the model's ability to combine objective and subjective factors in decision-making is unique. Therefore, OR course designers are called upon to introduce the model into the course contents in universities and colleges.

References

1. Aravindan, P., Punniyamoorthy, M.: Justification of advanced manufacturing technologies (AMT). Int. J. Adv. Manuf. Technol. **19**, 151–156 (2002)
2. Bagum, N., Rashed, A.A., Masud, A., Islam, Q.: Using Multi-criteria Analysis in Decision Making Regarding the Adoption of Wind Pump for Irrigation in Bangladesh. Rev. Gen. Manage. **15**, (2012)
3. Basuki, A., Cahyani, A.D.: Decision Making of Warehouse Location Selection Using Brown-Gibson Model. Adv. Sci. Lett. **23**, 12381–12384 (2017)
4. Bhattacharya, A., Sarkar, B., Mukherjee, S.: A new method for plant location selection: a holistic approach. Int. J. Ind. Eng.Theory Appl. Pract. **11**, 330–338 (2004)
5. Brown, P.A., Gibson, D.F.: A quantified model for facility site selection-application to a multiplant location problem. AIIE Trans. **4**, 1–10 (1972)
6. Brown, P.A.: Plant location: a quantified model for community and plant site selection. (1970)
7. Buffa, E.S., Sarin, R.K.: Modern Production/Operations Management. John Willey & Sons. Inc., ABD (1987)

8. Corum, A.: Facility location problem. (2006)
9. Dominic, P.D.D., Mahmood, A.K., Murugesh, V., Sridevi, P.: Multiattribute analysis of the offshore outsourcing location decision using a decision support system framework. Int. J. Bus. Inf. Syst. **3**, 445–463 (2008)
10. Facchinetti, G., Ricci, R.G., Muzzioli, S.: Note on ranking fuzzy triangular numbers. Int. J. Intell. Syst. **13**, 613–622 (1998)
11. Fasal, J.: Forced decisions for value. Product. Engineering **36**, 84–86 (1965)
12. Feridun, M., Korhan, O., Özakça, A.: Multi-attribute decision making: an application of the Brown-Gibson model of weighted evaluation. (2005)
13. Fessi, B.A., Benabdallah, S., Boudriga, N., Hamdi, M.: A multi-attribute decision model for intrusion response system. Inf. Sci. **270**, 237–254 (2014)
14. Hemalatha, M., Sridevi, P., Sivakumar, V.: Multiattribute analysis of the retail store location decision. J. Contemp. Res. Manage. **3**, 43–52 (2008)
15. Hemalatha, M., Sridevi, P., Sivakumar, V.: A Decision-Support System application in retail store location model: a case study of hypermarket in emerging markets. Int. J. Bus. Emerg Markets **3**, 158–176 (2011)
16. Ichikawa, K., Tamano, H.: Unsupervised qualitative scoring for binary item features. Data Sci. Eng. **5**, 317–330 (2020)
17. Kaboli, A., Arianezhad, M., Shahanaghi, K., Tavakoli, M.R.: A holistic approach based on MCDM for solving location problems. (2007)
18. Kaboli, A., Aryanezhad, M., Shahanaghi, K., Tavakkoli, M.R.: A combined fuzzy-ahp and goal programming model for location-allocation problems. (2008)
19. Kumar, A., Sah, B., Singh, A.R., Deng, Y., He, X., Kumar, P., Bansal, R.: A review of multi criteria decision making (MCDM) towards sustainable renewable energy development. Renew. Sustain. Energy Rev. **69**, 596–609 (2017)
20. Liou, T.-S., Wang, M.-J.J.: Ranking fuzzy numbers with integral value. Fuzzy Sets Syst. **50**, 247–255 (1992)
21. Maniraj, M., Pakkirisamy, V.: Justification of reconfigurable manufacturing systems selection using extended Brown-Gibson model and fuzzy TOPSIS. Int. J. Ind. Syst. Eng. **20**, 1–21 (2015)
22. Mateo, J.R.S.C.: Multi-criteria analysis. In: Multi Criteria Analysis in the Renewable Energy Industry, pp. 7–10. Springer. (2012)
23. Parameshwaran, R., Srinivasan, P.: An integrated closed-loop model for service performance management. Int. J. Serv. Oper. Manage. **4**, 34–55 (2008)
24. Parameshwaran, R., Srinivasan, P., Punniyamoorthy, M.: Modified closed loop model for service performance management. Int. J. Qual. Reliab. Manage. (2009)
25. Parthiban, P., Abdul Zubar, H., Ganesh, K., Nagarajan, S.: MASS: an analytical model for assessment of supply chain entity. Int. J. Oper. Res. **18**, 318–345 (2013)
26. Parthiban, P., Goh, M.: An integrated model for performance management of manufacturing units. Benchmarking Int. J. (2011)
27. Punniyamoorthy, M., Ragavan, P.V.: Justification of automatic storage and retrieval system (AS/RS) in a heavy engineering industry. Int. J. Adv. Manuf. Technol. **26**, 653–658 (2005)
28. Ragavan, P., Punniyamoorthy, M.: A strategic decision model for the justification of technology selection. Int. J. Adv. Manuf. Technol. **21**, 72–78 (2003)
29. Rahman, M.M., Rahman, M.S, & Rafiquzzaman, M.: Automobile location selection in Bangladesh. 324644808 (2016)
30. Raja, T.M., Stalin, N.: Design of an Evaluation Model for Mid-size Car Segment using a DSS Framework. Int. J. Appl. Manage. Technol. **257**, (2007)
31. Saaty, T.L.: Decision making with the analytic hierarchy process. Int. J. Serv. Sci. **1**, 83–98 (2008)
32. Sanders, G.L., Ghandforoush, P., Austin, L.M.: A model for the evaluation of computer software packages. Comput. Ind. Eng. **7**, 309–315 (1983)
33. Tabari, M., Kaboli, A., Aryanezhad, M.-B., Shahanaghi, K., Siadat, A.: A new method for location selection: a hybrid analysis. Appl. Math. Comput. **206**, 598–606 (2008)

34. Triantaphyllou, E.: Multi-criteria decision making methods. In: Multi-Criteria Decision Making Methods: A Comparative Study, pp. 5–21. Springer, (2000)
35. Velasquez, M., Hester, P.T.: An analysis of multi-criteria decision making methods. Int. J. Oper. Res. **10**, 56–66 (2013)
36. Yimen, N., Dagbasi, M.: Multi-attribute decision-making: applying a modified Brown-Gibson model and RETScreen software to the optimal location process of utility-scale photovoltaic plants. Processes **7**, 505 (2019)
37. Yimen, N., Hamandjoda, O., Meva'a, L., Ndzana, B., Nganhou, J.,: Analyzing of a photovoltaic/wind/biogas/pumped-hydro off-grid hybrid system for rural electrification in Sub-Saharan Africa—case study of Djoundé in Northern Cameroon. Energies **11**, 2644 (2018)
38. Yimen, N., Tchotang, T., Kanmogne, A., Abdelkhalikh Idriss, I., Musa, B., Aliyu, A., Okonkwo, E.C., Abba, S.I., Tata, D., Meva'a, L.: Optimal sizing and techno-economic analysis of hybrid renewable energy systems—a case study of a photovoltaic/wind/battery/diesel system in Fanisau. Northern Nigeria. Processes **8**, 1381 (2020)

A Grey Approach for the Computation of Interactions Between Two Groups of Irrelevant Variables of Decision Matrices

Shervin Zakeri, Naoufel Cheikhrouhou, Dimitri Konstantas, and Fereshteh Sattari Barabadi

Abstract In this chapter, we aim to find a mathematical solution to compute the impact between two irrelevant decision matrices in a complex decision-making problem using multiple-criteria decision-making (MCDM) methods. The existing MCDM methods merely provide solutions for the one-stage decision-making procedure and do not take other effective variables outside of the decision matrix into account, while in real-world processes, the decisions always impact by the variables where they appear to be irrelevant. To demonstrate our proposed approach, it is applied to a case of supplier selection and firm's strategies in which the interaction of selected strategies has been investigated on the selection of the best supplier. In order to handle the uncertainty that emerge during the process, this four-section approach is implemented as a grey framework and deals with grey Entropy, grey-TOPSIS, and the grey strategies interaction model. With comparison of rankings in computation with impact of selected strategies and without them, results indicated essentially the difference between these two cases.

Keywords Multi-criteria decision-making method · Irrelevance · Interaction · Grey-TOPSIS · SIM

Abbreviations

MCDM	Multi-criteria decision-making method
DM	Decision-maker

S. Zakeri (✉) · D. Konstantas
Geneva School of Economics and Management, University of Geneva, 1211 Geneva, Switzerland
e-mail: Shervin.zakeri@etu.unige.ch

S. Zakeri · N. Cheikhrouhou
Geneva School of Business Administration, University of Applied Sciences Western Switzerland, HES-SO, 1227 Geneva, Switzerland

F. S. Barabadi
Department of Public Administration, Financial Management, Islamic Azad University, Mashhad, Iran

A. J. Kulkarni (ed.), *Multiple Criteria Decision Making*, Studies in Systems, Decision and Control 407, https://doi.org/10.1007/978-981-16-7414-3_10

TOPSIS	The Technique for Order of Preference by Similarity to Ideal Solution
G-TOPSIS	Grey-TOPSIS
SIM	Strategies interaction model
SWOT	Strengths, Weaknesses, Opportunities, Threats
SO	Strengths Opportunities
WO	Weaknesses Opportunities
ST	Strengths Threats
WT	Weaknesses Threats
QSPM	Quantitative Strategic Planning Matrix
VP	Very Poor
P	Poor
MP	Medium Poor
F	Fair
MG	Medium Good
G	Good
VG	Very Good
GUV	The Grey Uncertainty Value

List of Symbols

$\otimes G_1$	A grey number
$\left[\underline{G_1}, \overline{G_1}\right]$	Grey interval
$\underline{G_1}$	Grey lower bound
$\overline{G_1}$	Grey upper bound
e	Entropy
w	Weight
S^{max}	Positive ideal alternative
S^{min}	Negative ideal alternative
γ_{oi}	The grey relation coefficient
C_i	The grade of grey relation
$\otimes P = \left[\underline{P_{ij}}, \overline{P_{ij}}\right]$	A normalized grey number
$N_D = \left[N_{\underline{G}_{ij}}, N_{\overline{G}_{ij}}\right]$	The normalized decision matrix
D	The decision matrix
$\updownarrow$	The distance between the elements of each cloud with lower and upper bound
ϱ	GUV
ξ	The coefficient of uncertainty/probability

1 Introduction

In order to find the most appropriate solutions, multi-criteria decision-making methods (MCDM) are the translation systems, which translate decision-making problems, from less complex such as the daily decision-making problems to advanced decision-making problems, to the mathematical algorithms. MCDM methods are developed to analyze alternatives against the criteria with the various algorithms to lead the decision-maker (DM) to the optimal solutions for the decision-making problems.

In general, MCDM methods are employed to handle MCDM problem with the selection of a suitable solution among alternatives concerning a variety of factors [1]. Eyvindson et al. [2] described MCDM techniques as the mathematical methods employed to find a best compromise solution based on judgments provided by stakeholders [3]. According to [4], MCDM methods consist of ranking alternatives, or selecting an appropriate alternative, with respect to several multiple, conflicting, and interactive criteria. To solve different decision-making problems, MCDM methods are designed into integrated method, which two or three MCDM methods constitute an integrated approach, or group decision-making methods [5–7]. MCDM methods have been applied to a wide range of problems such as Supply chain management [8], Energy [9], Transportation [10], Logistic [11], Agriculture and water resource management [12], Civil engineering and construction management [13], Strategic decision-making [9], and Strategic management [13, 14].

MCDM methods are designed to analyze a set of alternatives given by the problem against a set of criteria in order to offer the solutions; yet, in the real-world processes, the decisions are affected by multiple variables originating from the external forces from outside of the decision matrices that are constructed by the MCDM methods. These variables possibly seem irrelevant to the decision-making problem; however, their impact is hidden in the final result. This brings a serious shortage when the problem is observed through a holistic view. To illustrate how the aforementioned process functions, in this chapter, a mathematical framework is proposed to calculate the impact and interaction between two irrelevant decision matrices. In fact, as discussed in advance, there are many issues that affect the decision-making where they need to be identified while they are not considered in the conventional decision-making methods. For instance, in the supplier selection problem, the lack of consideration of the firm's strategies may cause the wrong selection when the supplier selection, as the part of operational strategies execution, needs to be in line with the firm's strategies.

Supplier selection is a typical MCDM activity [15]. According to [16], selection of the proper suppliers will reduce costs and provide high quality products. El Hiri et al. [17] defined selection of the most proper suppliers as a vital activity for elevating the result of a company's efforts to conserve its market position. Indeed, as stated by [18], one of the key issues in supply chain management is supplier selection and also finding the best supplier among several alternatives against various criteria, such as services, cost, and risks. As mentioned heretofore, in real world the selections which they upon DM's decision occur under uncertainty environment.

As mentioned heretofore, the supplier selection process highly depends on experts' assessments. Yet, the issue emerges when the firms' strategies are not considered by DMs or in general in the decision-making process to select the suppliers. Ignoring the strategies in the supplier selection process mainly causes the lack of a comprehensive approach to select a supplier, provisional supplier selection, and inappropriate selection of the supplier as an integral part of the operational strategies implementation. On the other hand, all environmental planning and management decisions are subject to a number of uncertainties ranging from complexities of natural systems, variable degrees of unpredictable randomness, frequent lack of sufficient data, and at times, the politicized and therefore variable interpretation of information [19]. The certain decision approaches have been applied on various studies in the field of supplier selection such as Abdel-Baset et al. [20, 21], while certain decisions addressing are based upon classical assumptions and always tend to be so in deterministic conditions [22].

In the real-world application, with emerge of vagueness, uncertainty, or imprecision in the solutions evaluation, the final output is not a crisp value, but rather a distribution, a fuzzy value or a numerical interval which is called the grey number. In this chapter, the grey system is exercised for the computation of the interactions. The Grey System was first introduced by Deng [23]. The grey system theory is found as a channel in order to materialize incomplete information of individuals, professionals, etc., into discrete data [24]. It is widely applied in various fields of research and projects such as systems analysis, data processing, modeling, and prediction, as well as in control and decision-making [25]. Deng [23, 26] developed the grey decision-making systems. The grey decision is made in the situation that the decision model has grey elements or the normal decision model and grey model are combined, and the key research is the scheme selection problem [27]. Nowadays, the grey systems theory is broadly applied to different decision-making problems to handle the uncertainty [28, 29]. To calculate the interaction between the firm's strategies, which have been derived from grey strategies interaction model, and the supplier selection, grey TOPSIS (G-TOPSIS) and Grey-Entropy are utilized in this chapter. Indeed, the objective of this chapter is to propose a solution for the problem of connection between two irrelevant decision matrices which have impact on each other in real-world problems.

The rest of this chapter is organized as follows: in the Sect. 2, the grey numbers and their operation are demonstrated; the methods which are used in this chapter are represented in the Sect. 3. In Sect. 4, the proposed methodology is illustrated; in the Sect. 5, the application and results are discussed. The comparisons and discussion are stated in Sect. 6; and finally, Sect. 7 is devoted to the conclusion and future research.

2 Grey Numbers and Operations

The grey information refers to the partial knowledge and incomplete information in a three-section information box including the complete and known information, the incomplete information and the unknown information, where they are cited as the white, grey, and black information categories, respectively [30]. The grey systems theory presents three categories of uncertainty comprising the white, grey, and black numbers in accordance with the level of information. The meaning of information in the category of grey is given in the following table (Table 1).

According to [31], the four possibilities of emergence of grey information is given in the following list:

(1) The information about elements is grey;
(2) The structural information is grey;
(3) The boundary information is grey;
(4) The behavior information of motion is grey.

Grey systems theory and its operations are founded on the grey numbers which play a vital role in the application of grey methods [32]. Limited between two lower and upper bounds, the exact value of grey number is unknown, yet, the range where the value is located is known [31]. In fact, grey numbers stand for such numbers that are not crisp values, but some incomplete information [33]. Furthermore, Darvishi et al. [34] defined a grey number as a number with clear upper and lower boundaries, but which has an unclear position within the boundaries. The following equations (Eqs. 1–10) address the grey number operations:

$$If \otimes G_1 = \left[\underline{G_1}, \overline{G_1}\right], \otimes G_2 = \left[\underline{G_2}, \overline{G_2}\right]\ then\ \overline{G_1} > \underline{G_1}\ and\ \overline{G_2} > \underline{G_2} \quad (1)$$

$$-\otimes G_1 = \left[-\overline{G_1}, -\underline{G_1}\right] \quad (2)$$

$$\otimes G_1 + \otimes G_2 = \left[\underline{G_1} + \underline{G_2}, \overline{G_1} + \overline{G_2}\right] \quad (3)$$

$$\otimes G_1 - \otimes G_2 = \otimes G_1 + (-\otimes G_2) = \left[\underline{G_1} - \overline{G_2}, \overline{G_1} - \underline{G_2}\right] \quad (4)$$

$$\otimes G_1 \times \otimes G_2 = \left[min\left\{\underline{G_1}\underline{G_2}, \underline{G_1}\overline{G_2}, \overline{G_1}\underline{G_2}, \overline{G_1}\overline{G_2}\right\}, \max\left\{\underline{G_1}\underline{G_2}, \underline{G_1}\overline{G_2}, \overline{G_1}\underline{G_2}, \overline{G_1}\overline{G_2}\right\}\right] \quad (5)$$

Table 1 The information meaning of the grey

	Information	Appearance	Process	Property	Methodology	Attitude	Conclusion
The grey	Incomplete	Grey	Replace old with new	Complexity	Transition	Tolerance	Multiple solutions

$$r \times \otimes G_1 = \left[r\underline{G_1}, r\overline{G_1}\right] \tag{6}$$

$$\begin{aligned} {}^{\otimes G_1}/_{\otimes G_2} = & \left[\underline{G_1}, \overline{G_1}\right] \times \left[\frac{1}{\underline{G_2}}, \frac{1}{\overline{G_2}}\right] = \left[\underline{G_1}, \overline{G_1}\right] \times \left[\underline{G_2^{-1}}, \overline{G_2^{-1}}\right] \\ = & \left[\min\left\{\underline{G_1}\underline{G_2^{-1}}, \underline{G_1}\overline{G_2^{-1}}, \overline{G_1}\overline{G^{-1}}\right\}, \max\left\{\underline{G_1}\underline{G_2^{-1}}, \underline{G_1}\overline{G_2^{-1}}, \overline{G_1}\overline{G^{-1}}\right\}\right] \end{aligned} \tag{7}$$

$$\frac{\otimes G_1}{a} = \left[\frac{\underline{G_1}}{a}, \frac{\overline{G_1}}{a}\right] \tag{8}$$

$$\frac{a}{\otimes G_1} = \left[\frac{a}{\underline{G_1}}, \frac{a}{\overline{G_1}}\right] \tag{9}$$

The possibility degree of $\otimes G_1 \le \otimes G_2$:

$$p\{\otimes G_1 \le \otimes G_2\} = \frac{\max(0, L^* - \max(0, \underline{G_1} - \overline{G_2}))}{L^*} \tag{10}$$

where $L^* = L(\otimes G_1) + L(\otimes G_2)$.

3 Methods and Tools

The proposed approach has been applied on a strategic supplier selection problem where the suppliers are selected in accordance with the firm's strategies. To run the approach, the grey Shannon's Entropy and grey TOPSIS are employed. In this section, these two methods of algorithms are discussed.

3.1 Strategies Interaction Model (SIM)

Strengths, weaknesses, opportunities, and threats (SWOT) analysis is a management tool to formulate strategic action plans [35]. As a strategic management tool, SWOT analysis has been extensively utilized for the decision-making process [36]. According to Gao and Peng [37], SWOT analysis is an important decision-making support tool, and is commonly used to systematically analyze the strategic situations and identify the level of organizations from their internal and external environments. SWOT matrix analyzes the internal strengths and weaknesses as well as external opportunities and threats to derive promising future strategies [38]. It also prioritizes the strategies by the Quantitative Strategic Planning Matrix (QSPM) in the classic form. However, due to the flexible structure of its approach and the fact that SWOT provides only a qualitative analysis that merely prioritizes the factors' importance by measuring them quantitatively, thus, fails to address the rank of the strategies, hence, mostly, it integrates other decision-making methods such as MCDM techniques [39].

Application and integration of MCDM methods with SWOT analysis process could be addressed in different studies such as Anser et al. [40, 41]. The strategies derived from SWOT matrix are categorized into four groups of SO strategies where they use strengths to take advantage of opportunities, WO strategies where they overcome weaknesses by taking advantage of opportunities, ST strategies in which they use strengths to avoid threats, and WT strategies which minimize weaknesses to avoid threats [42], likewise, these strategies are addressed as the aggressive strategies, competitive strategies, conservative strategies, and the defensive strategies.

The classic form of SWOT itself and its integration with MCDM methods is suffering from a number of shortages including [43]:

1. Ignoring the strategic position ignorance in MCDM and SWOT integrated methodologies.
2. Lack of an integrated model for the selection of an organization strategies and also alternative strategies in order to the organization strategic position.
3. In spite of the shared resources for the execution of strategies, there is no framework to assess the interaction of strategies due to their budget requirement.
4. Lack of a formulated paradigm to support the assessment of the interaction of the possible unselected strategies on the main selected strategies ranking.

To cover the aforementioned lacks through the classic SWOT analysis process, Zakeri et al. [43] proposed an approach to analyze SWOT, called strategies interaction model which is divided into two main areas: the evaluations area where the evaluation and all computation activities are executed, and the selection area in which the results are processed (see Fig. 1).

SIM are designed in the grey environment. According to [43], the SIM phases are as follows:

Phase I. Analysis of internal and external factors.
Phase II. Construction of SWOT matrix.

Phase II.I. Selection of the strategies (All strategic positions).
Phase II.II. Determination of strategic position and selection of the strategies in accordance with the strategic position.

Phase III. Computation of the value of interaction.
Phase IV. Ranking of the selected strategies.
Phase V. Evaluation and selection of the alternative strategies.

3.2 Shannon's Entropy

One of the major results of information theory is the Shannon entropy [28], Shannon [44]. This method has been utilized to compute the weights of the criteria in a decision-making problem. With respect to [45], the grey entropy is in accordance with (Eqs. 11 and 12).

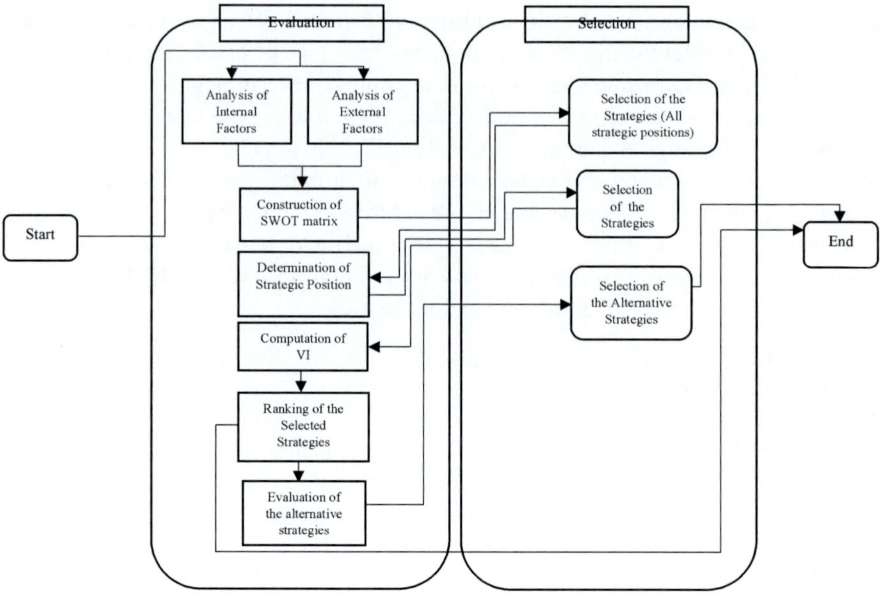

Fig. 1 The proposed methodology procedure of SIM

$$e_{\underline{G}_j} = -\frac{1}{\ln m}\sum_{i=1}^{m}\underline{G}_{ij}\ln\underline{G}_{ij} \tag{11}$$

$$e_{\overline{G}_j} = -\frac{1}{\ln m}\sum_{i=1}^{m}\overline{G}_{ij}\ln\overline{G}_{ij} \tag{12}$$

The weight of Jth criterion is computed by following Eqs. (13 and 14):

$$w_{\underline{G}_j} = \left(1 - e_{\underline{G}_j}\right).\left(\sum_{j=1}^{n}(1 - e_{\underline{G}_j})\right)^{-1} \tag{13}$$

$$w_{\overline{G}_j} = (1 - e_{\overline{G}_j}).\left(\sum_{j=1}^{n}(1 - e_{\overline{G}_j})\right)^{-1} \tag{14}$$

3.3 Grey TOPSIS

One of the most popular MCDM technique which is widely applied to solve MCDM problems is TOPSIS. Hwang and Yoon [46] first proposed a technique for establishing order performance by referencing its similarity to the ideal solution (TOPSIS). The TOPSIS philosophy is that the selected alternative's value should have the shortest distance from the ideal solution and the farthest distance from the negative ideal solution [47]. The grey TOPSIS has the following steps [22, 48]:

Step.3.2.1. Constructing the decision matrix.

Step.3.2.2. Establishing the normalized decision matrices with respect to the cost or benefit (Eqs. 15 and 16).

For benefit attribute of $\otimes G_{ij}^{*}$, the normalization is defined as in the following equation:

$$\otimes G_{ij}^{+} = \left[\frac{\underline{G_{ij}}}{G_j^{max}}, \frac{\overline{G_{ij}}}{G_j^{max}} \right] \tag{15}$$

where $\otimes G_{ij} = \left[\underline{G_{ij}}, \overline{G_{ij}} \right]$ and $\otimes G_j^{max} = \max\limits_{1 \le i \le m} \left\{ \overline{G_{ij}} \right\}$.

And for a cost attribute of $\otimes G_{ij}^{*}$, there is the following equation:

$$\otimes G_{ij}^{-} = \left[\frac{\underline{G_j^{min}}}{\overline{G_{ij}}}, \frac{G_j^{min}}{\underline{G_{ij}}} \right] \tag{16}$$

where $\otimes G_j^{max} = \min\limits_{1 \le i \le m} \left\{ \underline{G_{ij}} \right\}$.

Step 3.2.3. Construction of the weighted normalized matrix.

Step 3.2.4. Calculation of (S^{max}) as the ideal alternative where (S^{max}) is a referential alternative (Eq. 17).

$$\begin{cases} S^{max} = \left\{ G_1^{max}, G_2^{max}, G_3^{max}, \ldots, G_n^{max} \right\}; \\ S^{max} = \left\{ \left[\max\limits_{1 \le i \le m} \underline{V_{i1}}, \max\limits_{1 \le i \le m} \overline{V_{i1}} \right], \ldots, \left[\max\limits_{1 \le i \le m} \underline{V_{in}}, \max\limits_{1 \le i \le m} \overline{V_{in}} \right] \right\}; \end{cases} \tag{17}$$

Step 3.2.5. Computation of the distance between each of the alternatives' sequences (Eq. 18).

$$d = \Delta_{\otimes G_1 - \otimes G_2} = \left(\underline{G_1} - \underline{G_2} \right) + \left(\overline{G_1} - \overline{G_2} \right) \tag{18}$$

where Δ is the distance d between two grey numbers of $\otimes G_1$ and $\otimes G_2$.

Step 3.2.6. Determination of the grey relation coefficient between each of the alternatives (Eq. 19)

$$\gamma_{oi} = \gamma(x_o(j), x_i(j)) = \frac{\underset{i}{\min}\underset{j}{\min} d_{ij} + \xi \underset{i}{\max}\underset{j}{\max} d_{ij}}{d_{ij} + \xi \underset{i}{\max}\underset{j}{\max} d_{ij}} \tag{19}$$

Step 3.2.7. Computing the grade of grey relation of each alternative to the ideal solution in accordance with the following equation (Eq. 20):

$$C_i = \left(1 - \frac{1}{n} \cdot sum_{j=1}^{n} \gamma_{ij}\right) i = 0, 1, \ldots, m \tag{20}$$

Step 3.2.8. The final step is the prioritization of the alternatives according to the higher score of C_i.

4 Proposed Methodology

In this chapter, to address the proposed approach, a computation of interaction between a firm's strategies and its supplier selection problem is presented in order to have a supplier selection in line with the firm's strategies. Various studies have employed SWOT analysis for the supplier selection [5, 49], while none of them did not exercise SWOT analysis for the specific reason of alignment of the supplier selection with the firm's strategies. The implementation of the proposed approach is designed in four steps including:

Step 1. Selecting the firm's strategies through the grey strategies' interaction model (G-SIM).

Step 2. Evaluation of the criteria. The main purpose of this section is calculating of (λ). Indeed, (λ) is the proposed method's key element. In this step, the strategies derived from the first section are playing the role of the criteria and the main criteria of supplier selection are the alternatives in a decision matrix.

Step 3. Prioritization of criteria is based on relation matrix. In this step, the (λ) is determined by the normalized performance of each criteria ranking. The strategies directly impact on supplier selection by (λ).

Step 4. The final section is selection of the best supplier.

The proposed method's steps are illustrated in (Fig. 2).

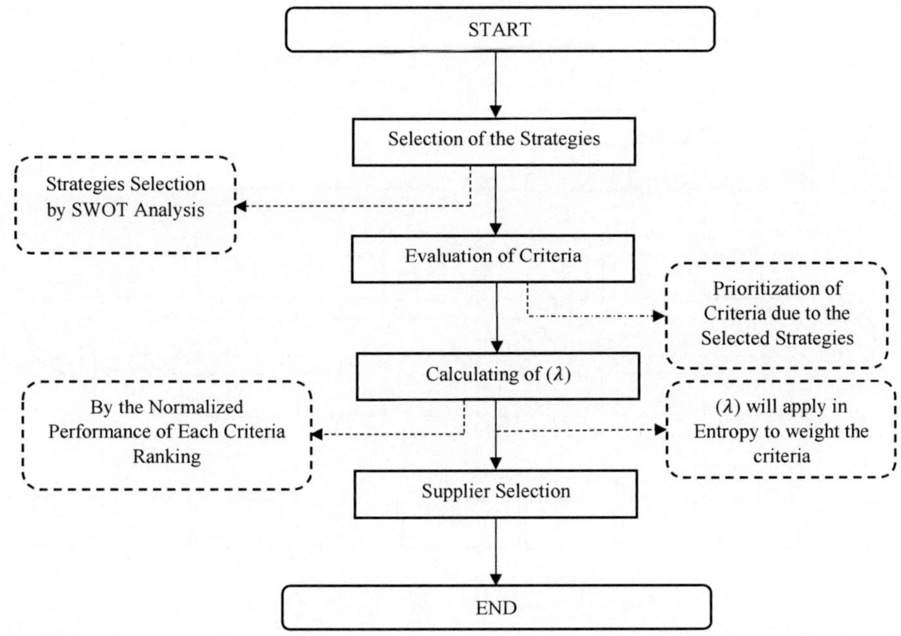

Fig. 2 The proposed method's chart

5 Method Application and Results

In this section, the proposed model is represented as a numerical example. This section is separated into three main parts including: 1. Selection of the strategies; 2. Evaluation of Criteria: Prioritization of Criteria based on the Selected Strategies; 3. Supplier selection.

5.1 Selection of Strategies SIM

SWOT analysis of a firm is illustrated (Fig. 3). Selected strategies have been carried out by implementation of SIM. In the following figure, SO, ST, WO, WT stand for the aggressive, competitive, conservative, and defensive strategies.

According to the scores, S_1O_1, S_2O_1, $S_1T_{1,2}$, W_1O_2, W_2T_3 are selected as the best strategies. These strategies are shown as the ST_1, ST_2, ST_3, ST_4, and ST_5 in next steps of the proposed method application.

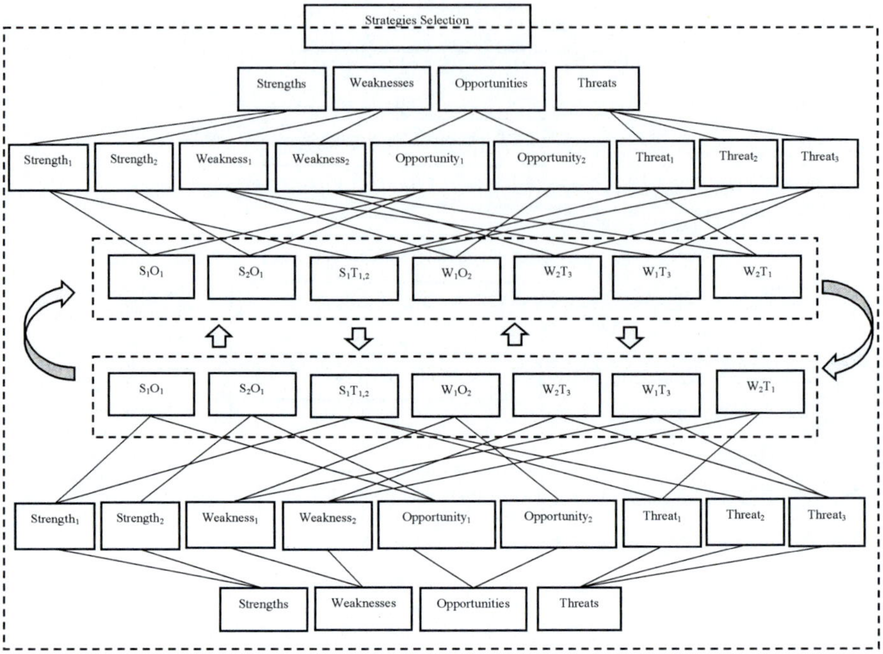

Fig. 3 SIM structure of SWOT analysis for ranking and selection of the strategies

5.2 Evaluation of Criteria: Prioritization of Criteria Due to the Selected Strategies

In this section, the criteria for the supplier selection are prioritized in accordance with the selected strategies derived from the previous section. To rank the criteria, the relation matrix has been utilized. In the relation matrix, the interaction of the variables is investigated through computation of their relationship. With respect to the relation matrix, the degree of the relationship between alternatives and criteria are evaluated by the linguistic variables. In this step, the prioritization process is performed by the G-TOPSIS algorithm. The grey linguistic variables are presented in Table 2.

The following tables (Tables 3 and 4) demonstrate the relation matrix, where the supplier evaluation criteria are the alternatives, and the selected strategies are the criteria.

With respect to the Eqs. (11–13), in most grey-based MCDM problems, for objective calculation of the weights of criteria, the grey entropy is employed. In this chapter, a novel form of grey entropy algorithm is designed to transform the grey numbers to white numbers (Eqs. 30–33); indeed, the new algorithm is in line with the grey entropy which is proposed by [45] with respect to the Eqs. (11–13).

If

Table 2 The grey attributes scale of rating of $\otimes G$

Scale	Grey
VP	[0, 1]
P	[1, 3]
MP	[3, 4]
F	[4, 5]
MG	[5, 6]
G	[6, 9]
VG	[9, 10]

Table 3 The relation matrix between selected strategies and supplier selection criteria with linguistic variables

	ST_1	ST_2	ST_3	ST_4	ST_5
C_1	F	P	G	MG	MP
C_2	MG	P	VG	G	F
C_3	MG	MP	VG	F	MP
C_4	F	F	G	F	P
C_5	G	P	F	MG	G

Table 4 The relation matrix between selected strategies and supplier selection criteria

	ST_1	ST_2	ST_3	ST_4	ST_5
C_1	[4, 5]	[1, 3]	[6, 9]	[5, 6]	[3, 4]
C_2	[5, 6]	[1, 3]	[9, 10]	[6, 9]	[4, 5]
C_3	[5, 6]	[3, 4]	[9, 10]	[4, 5]	[3, 4]
C_4	[4, 5]	[4, 5]	[6, 9]	[4, 5]	[1, 3]
C_5	[6, 9]	[1, 3]	[4, 5]	[5, 6]	[6, 9]

$$\Gamma \geq \left(\underline{G}_{ij} + \overline{G}_{ij}\right) \tag{21}$$

And

$$\Phi \geq \left(\underline{G}_{ij} - \overline{G}_{ij}\right) \tag{22}$$

Then:

$$e_{\otimes G_j} \geq \left[\sum_{i=1}^{m}\left(\left(\sum_{i=1}^{m}((\Phi^2)\ln(\Phi^2))^2\right)^{\frac{1}{4}} \cdot \left(\left(\sum_{i=1}^{m}((\Gamma^2)\ln(\Gamma^2))^2\right)^{\frac{1}{4}}\right)^{-1}\right)\right](\ln m)^{-1} \tag{23}$$

$$w_{\otimes G_j} \geq \left(\sum_{j=1}^{n}(1 - e_{\otimes G_j})\right)^{-1} .(1 - e_{\otimes G_j}) \tag{24}$$

The first step of Entropy is the normalization of the decision matrix. The normalization process is as given in (Eqs. 25 and 26):

If $\otimes P = \left[\underline{P_{ij}}, \overline{P_{ij}}\right]$ stands for a normalized grey number of $\otimes G = \left[\underline{G_{ij}}, \overline{G_{ij}}\right]$ in a decision matrix, therefore

$$\underline{P_{ij}} = \left((\sum_{i=1}^{m}(\underline{G}_{ij})^2)^{\frac{1}{2}}\right)^{-1} .\underline{G}_{ij} \tag{25}$$

$$\overline{P_{ij}} = \left((\sum_{i=1}^{m}(\overline{G}_{ij})^2)^{\frac{1}{2}}\right)^{-1} .\overline{G}_{ij} \tag{26}$$

With the following transportation of grey numbers to white numbers, introduced in (Eqs. 21 and 22), another transportation of the normalization process is proposed in this chapter which can be found in (Eq. 27), where (P) is a crisp number and white number of $\otimes G = \left[\underline{G_{ij}}, \overline{G}_{ij}\right]$. Yet, the proposed framework of this chapter deals with the original normalization processes in accordance with Eqs. (25 and 26).

$$P \geq \left(\prod_{i=1}^{m}(\sum_{i=1}^{m}\underline{G_{ij}} + \overline{G}_{ij})\right)^{-1} .\left(\underline{G_{ij}} + \underline{G_{ij}}\right) \tag{27}$$

ccording to Eqs. (25 and 26), the normalized decision matrix is displayed in Table 5.

Weights of each selected strategies in the relation matrix can be found in Table 6.

For calculation of the $(e^{'}_{j})$, the process followed is given in Eq. (28). As pictured in Table 6, there are some anomalies for the normalized interval of each strategies; in other words, in some intervals, lower bound is larger than upper bound. To overcome this problem, for the calculation of the weight of each strategy, we have proposed the following equation (Eq. 28).

Table 5 The normalized relation matrix

	ST_1	ST_2	ST_3	ST_4	ST_5
C_1	[0.368, 0.351]	[0.189, 0.364]	[0.379, 0.457]	[0.460, 0.421]	[0.356, 0.330]
C_2	[0.460, 0.421]	[0.189, 0.364]	[0.569, 0.508]	[0.552, 0.631]	[0.474, 0.412]
C_3	[0.460, 0.421]	[0.562, 0.485]	[0.569, 0.508]	[0.368, 0.351]	[0.356, 0.330]
C_4	[0.368, 0.351]	[0.756, 0.606]	[0.379, 0.457]	[0.368, 0.351]	[0.118, 0.247]
C_5	[0.552, 0.631]	[0.189, 0.364]	[0.253, 0.254]	[0.460, 0.421]	[0.712, 0.742]

Table 6 Entropy and weight of each strategies

	ST_1		ST_2		ST_3		ST_4		ST_5	
e_j	1.111	1.089	0.919	1.092	1.071	1.088	1.105	1.089	0.983	1.034
$e_j^{'}$	0.50005		0.5037		0.50003		0.50002		0.50031	
d_j	0.49995		0.4963		0.49997		0.49998		0.49969	
w_j	0.20030		0.1988		0.20031		0.20032		0.20020	

If $e_j = \left[\underline{G}, \overline{G}\right]$,then:

$$e_j^{'} = \left((\underline{G} + \overline{G})^2\right)^{-1} \cdot \left(\underline{G}^2 + \overline{G}^2\right) \tag{28}$$

In this chapter, to rank the criteria of supplier selection, we have used the transformation methodology proposed by [43]. The method is developed from the weighted product model (WPM)'s procedure. The proposed methodology could be found in Eqs. (29–31), where $\otimes G = \left[\underline{G}_{ij}, \overline{G}_{ij}\right]$ and $\otimes G^{'} = \left[\underline{G}_{ij}^{'}, \overline{G}_{ij}^{'}\right]$:

$$\underline{G}_{ij}^{'} = \left(\sum_{i=1}^{m} \underline{G}_{ij}\right)^{-1} \cdot \underline{G}_{ij} \tag{29}$$

$$\overline{G}_{ij}^{'} = \left(\sum_{i=1}^{m} \overline{G}_{ij}\right)^{-1} \cdot \overline{G}_{ij} \tag{30}$$

$$P\left(G^{'}_{m}\right) = \prod_{j=1}^{n} \left(\underline{G}_{ij} + \overline{G}_{ij}\right)^{w_j} \tag{31}$$

The prioritization is based on the larger value of $P\left(G^{'}_{m}\right)$, thus, with respect to the Eqs. (29–31), the normalized relation matrix is shown in Table 7, and (λ) values are displayed in Table 8.

Table 7 The normalized relation matrix

w_j	0.20030	0.1988	0.20031	0.20032	0.20020
	ST_1	ST_2	ST_3	ST_4	ST_5
C_1	[0.166, 0.161]	[0.100, 0.166]	[0.176, 0.209]	[0.208, 0.193]	[0.176, 0.160]
C_2	[0.208, 0.193]	[0.100, 0.166]	[0.264, 0.232]	[0.250, 0.290]	[0.235, 0.200]
C_3	[0.208, 0.193]	[0.300, 0.222]	[0.264, 0.232]	[0.166, 0.161]	[0.176, 0.160]
C_4	[0.166, 0.161]	[0.400, 0.277]	[0.176, 0.209]	[0.166, 0.161]	[0.588, 0.120]
C_5	[0.250, 0.290]	[0.100, 0.166]	[0.117, 0.116]	[0.208, 0.193]	[0.353, 0.360]

Table 8 Value of (λ) for each criteria

	C_1	C_2	C_3	C_4	C_5
λ'_m	0.666	0.866	0.8	0.933	0.733
λ_m	0.1665	0.2165	0.2	0.23325	0.18325

Therefore, according to Eq. (31), the larger value of $P\left(G'_m\right)$ possesses the best rank:

$$P\left(G'_1\right) = 4.030480387802214041019126058824 3$$

$$P\left(G'_2\right) = 4.2006219001599303350681892865139$$

$$P\left(G'_3\right) = 4.1836985480191150780837110210309$$

$$P\left(G'_4\right) = 4.2833361779410137110629886275380$$

$$P\left(G'_5\right) = 4.1665975424376581947939114823975$$

Hence, the ranking is as follows:

$$C_4 > C_2 > C_3 > C_5 > C_1$$

As mentioned heretofore, the next step of the proposed methodology is calculation of (λ). To calculate (λ), the number of each criteria ranking will be normalized by the normalized performance method. In this chapter, Eqs. (31 and 32) are employed to compute (λ).

$$\lambda'_m = 1 - \left(R_m \cdot \left(\sum_m^1 R_m\right)^{-1}\right) \tag{32}$$

$$\lambda_m = \lambda'_m \cdot \left(\sum_m^1 \lambda'_m\right)^{-1} \tag{33}$$

where (R_m) is the ranking of mth alternative, therefore, (λ) of each criteria is.

5.3 Supplier Selection

The final step of the proposed approach is the selection of the best supplier. In this chapter, Grey-TOPSIS is utilized for the supplier selection procedure. The classic Grey-TOPSIS algorithm is as followed in Eqs. (14–20), while in this chapter, we have proposed a novel algorithm for Grey-TOPSIS.

The following steps and equations express the new process of Grey-TOPSIS algorithm.

Step 5.3.1. Construction of Normalized Decision Matrix

$$N_D = \left[N_{\underline{G}_{ij}}, N_{\overline{G}_{ij}}\right] \tag{34}$$

$$N_D = \begin{cases} N_{\underline{G}_{ij}} = \left(\sum_{i=1}^{m} \underline{G}_{ij}\right)^{-1} . \underline{G}_{ij} \\ N_{\overline{G}_{ij}} = \left(\sum_{i=1}^{m} \overline{G}_{ij}\right)^{-1} . \overline{G}_{ij} \end{cases} \tag{35}$$

where D denotes the decision matrix and N_D stands for the normalized decision matrix of D.

Step 5.3.2. Establishing Weighted Normalized Decision Matrix

$$V = W \times N_D \tag{36}$$

where V states weighted normalized matrix.

Step 5.3.3. Calculation of Positive and Negative Ideal Solution

$$S^{max} = \{G_1^{max}, G_2^{max}, G_3^{max}, \ldots, G_n^{max}\}; \tag{37}$$

$$S^{max} = \left\{\left[\max_{1 \le i \le m} \underline{V_{i1}}, \max_{1 \le i \le m} \overline{V_{i1}}\right], \ldots, \left[\max_{1 \le i \le m} \underline{V_{in}}, \max_{1 \le i \le m} \overline{V_{in}}\right]\right\}; \tag{38}$$

$$S^{min} = \{G_1^{min}, G_2^{min}, G_3^{min}, \ldots, G_n^{min}\}; \tag{39}$$

$$S^{min} = \left\{\left[\min_{1 \le i \le m} \underline{V_{i1}}, \min_{1 \le i \le m} \overline{V_{i1}}\right], \ldots, \left[\min_{1 \le i \le m} \underline{V_{in}}, \min_{1 \le i \le m} \overline{V_{in}}\right]\right\}; \tag{40}$$

To calculate S^{max} and S^{min}, we have proposed the following equation:

$$\omega_{ij} = \left(\underline{G}_{ij} + \overline{G}_{ij}\right)^{W_j} \tag{41}$$

Larger value of (ω_{ij}) is S^{max} and the smaller value is S^{min}.

Step 5.3.4. Prioritization of Alternatives

$$\gamma_i = \left(\xi.\left(\sum_{j=1}^{n}(V_{ij}-V_{\overline{ij}})^2\right)^{0.5}\right).\left(\left(\xi.\left(\sum_{j=1}^{n}(V_{ij}-V_{\overline{ij}})^2\right)^{0.5}\right) + \left(\sum_{j=1}^{n}(V_{ij}-V_{\overline{ij}})^2\right)^{0.5}\right)^{-1} \quad (42)$$

where γ_i is larger, the ranking order of alternative is better. Otherwise, the ranking order is worse. To implement the proposed developed Grey-TOPSIS algorithm, first, the decision matrix needs to be normalized. The decision matrix has been expressed in Tables 9 and 10 in which $C_j = \{C_1, ..., C_5\}$ is the set of criteria.

Next step is the normalization of the decision matrix with respect to the Eqs. (34 and 35). The normalized decision matrix is demonstrated in Table 11.

To calculate the weighted normalized decision matrix, weight of each criterion needs to be computed. Indeed, the key of the proposed approach appears in this step.

Table 9 Supplier selection decision-making matrix with the attributes scale of rating

	C_1	C_2	C_3	C_4	C_5
A_1	*MG*	*MP*	*F*	*P*	*MG*
A_2	*F*	*G*	*F*	*F*	*F*
A_3	*VG*	*F*	*VG*	*MP*	*MP*
A_4	*G*	*P*	*G*	*MG*	*P*

Table 10 Supplier selection decision-making matrix

	C_1	C_2	C_3	C_4	C_5
A_1	[5, 6]	[3, 4]	[4, 5]	[1, 3]	[5, 6]
A_1	[4, 5]	[6, 9]	[4, 5]	[4, 5]	[4, 5]
A_1	[9, 10]	[4, 5]	[9, 10]	[3, 4]	[3, 4]
A_1	[6, 9]	[1, 3]	[6, 9]	[5, 6]	[1, 3]

Table 11 The normalized supplier selection decision matrix

	C_1	C_2	C_3	C_4	C_5
A_1	[0.208, 0.200]	[0.214, 0.190]	[0.174, 0.172]	[0.077, 0.166]	[0.384, 0.333]
A_2	[0.166, 0.166]	[0.428, 0.428]	[0.174, 0.172]	[0.308, 0.222]	[0.308, 0.277]
A_3	[0.375, 0.333]	[0.286, 0.238]	[0.391, 0.345]	[0.231, 0.222]	[0.231, 0.222]
A_4	[0.250, 0.300]	[0.071, 0.142]	[0.261, 0.310]	[0.384, 0.333]	[0.077, 0.166]

For the computation of the weights of criteria, we utilized Entropy in accordance with Eqs. (11–13), while to calculate the impact of (λ), the chapter deals with the following equations:

$$w_{\underline{G}_j} = \lambda_n\left(1 - \left(-\frac{1}{\ln m}\sum_{i=1}^{m}\underline{G}_{ij}\ln\underline{G}_{ij}\right)\right)\left(\sum_{j=1}^{n}\lambda_n\left(-\frac{1}{\ln m}\sum_{i=1}^{m}\underline{G}_{ij}\ln\underline{G}_{ij}\right)\right) \tag{43}$$

$$w_{\overline{G}_j} = \lambda_n\left(1 - \left(-\frac{1}{\ln m}\sum_{i=1}^{m}\overline{G}_{ij}\ln\overline{G}_{ij}\right)\right)\left(\sum_{j=1}^{n}\lambda_n\left(-\frac{1}{\ln m}\sum_{i=1}^{m}\overline{G}_{ij}\ln\overline{G}_{ij}\right)\right) \tag{44}$$

As the interaction between supplier selection criteria and the selected strategies, (λ) impact on the suppliers' prioritization. In real-world problems, there are other elements which impact on supplier prioritization and increase complexity of selection. In this chapter, we also have added DM's decision as the weight of criteria other than weights which are derived from Entropy's equations and (λ). To apply DM's decision, the chapter deals with other proposed equations as follows:

$$W'_{\underline{G}_j} = \left(w_{\underline{G}_j}.w_{\underline{G}_{jDM}}\right).\left(\sum_{J=1}^{n} w_{\underline{G}_j}.w_{\underline{G}_{jDM}}\right)^{-1} \tag{45}$$

$$W'_{\overline{G}_j} = \left(w_{\overline{G}_j}.w_{\overline{G}_{jDM}}\right).\left(\sum_{J=1}^{n} w_{\overline{G}_j}.w_{\overline{G}_{jDM}}\right)^{-1} \tag{46}$$

where, as DM's decision, $\otimes w_{G_{jDM}} = [w_{\underline{G}_{jDM}}, w_{\overline{G}_{jDM}}]$ is a grey numerical interval number. However, if DM's decision is a crisp number, the process needs to follow the application of Eqs. (47 and 48). In this equation, (Eq. 6) procedure is also exercised.

$$W'_{\underline{G}_j} = \left(w_{\underline{G}_j}.w_{DM}\right).\left(\sum_{J=1}^{n} w_{\underline{G}_j}.w_{DM}\right)^{-1} \tag{47}$$

$$W'_{\overline{G}_j} = \left(w_{\overline{G}_j}.w_{DM}\right).\left(\sum_{J=1}^{n} w_{\overline{G}_j}.w_{DM}\right)^{-1} \tag{48}$$

The weights (by DM's decision), (λ), derived weights from Entropy algorithm, and the final weights have been exposed in Table 12 where DM's decisions are in the form of the grey numbers and calculation of e_j is in accordance with Eq. (29).

The weighted normalized matrix with respect to Eq. (38) is displayed in Table 13.

In this paper, to find positive and negative ideal solutions (Eqs. 37–41), we proposed a methodology to calculate the (GUV) of each interval where (δ) stands for GUV. The larger value of (δ) in each column of decision matrix is the (S^{max}),

Table 12 Normalized supplier selection decision matrix and weights of criteria

λ_j	0.1665	0.2165	0.2	0.23325	0.18325
e_j	0.500001	0.511	0.500006	0.5001	0.5005
d_j	0.499999	0.489	0.499994	0.4999	0.4995
$W_{\otimes G_j}$	0.1674	0.2129	0.2011	0.2345	0.1841
W_{DM}	[0.15, 0.15]	[0.25, 0.25]	[0.30, 0.30]	[0.175, 0.175]	[0.125, 0.125]
$W'_{\otimes wG_{jDM}}$	[0.1238, 0.1238]	[0.2625, 0.2625]	[0.2976, 0.2976]	[0.2024, 0.2024]	[0.1135, 0.1135]
	C_1	C_2	C_3	C_4	C_5
A_1	[0.208, 0.200]	[0.214, 0.190]	[0.174, 0.172]	[0.077, 0.166]	[0.384, 0.333]
A_2	[0.166, 0.166]	[0.428, 0.428]	[0.174, 0.172]	[0.308, 0.222]	[0.308, 0.277]
A_3	[0.375, 0.333]	[0.286, 0.238]	[0.391, 0.345]	[0.231, 0.222]	[0.231, 0.222]
A_4	[0.250, 0.300]	[0.071, 0.142]	[0.261, 0.310]	[0.384, 0.333]	[0.077, 0.166]

Table 13 The weighted normalized matrix

	C_1	C_2	C_3	C_4	C_5
A_1	[0.0257, 0.0247]	[0.0561, 0.0498]	[0.0516, 0.0512]	[0.0155, 0.0336]	[0.0436, 0.0378]
A_2	[0.0205, 0.0205]	[0.1123, 0.1123]	[0.0516, 0.0512]	[0.0623, 0.0449]	[0.0349, 0.0314]
A_3	[0.0464, 0.0412]	[0.0751, 0.0624]	[0.1163, 0.1027]	[0.0467, 0.0449]	[0.0262, 0.0252]
A_4	[0.0309, 0.0371]	[0.0179, 0.0358]	[0.0777, 0.0922]	[0.0777, 0.0674]	[0.0087, 0.0188]

otherwise it is (S^{min}). The following algorithm shows the steps of the computation of (δ).

Step 5.4.1. First step of the algorithm is making a cloud of number for each number in decision matrix. The cloud includes the set of ($\propto_n$) where $\propto_n = \{1, 2, ..., 9\}$. The elements are the set of numbers which are closest to the zero in the weighted decision matrix.

Step 5.4.2. Making another cloud of another set of ($\propto_n$), which includes ($\propto'_1$, $\propto'_2$, ..., $\propto'_n$). In this proposed methodology, it is assumed that two clouds by default (at least), but, if it is more than two zero in the first numbers, creating the clouds will continue to the first number. For instance, in ($0.\underline{0027}$) there are three clouds, while in ($0.\underline{00027}$) there are four clouds, yet, for ($0.\underline{02}73$) and ($0.\underline{27}34$) there are two clouds.

Step 5.4.3. Calculating distance between the elements of each cloud with lower and upper bound with respect to Eqs. (51–52).

$$\underline{\updownarrow} = \left(\sum_{j=1}^{n}\left(\underline{G_{ij}} - \propto_n\right)^2\right)^{0.5}, n = 1, 2, ..., 9; \tag{49}$$

$$\overline{\updownarrow} = \left(\sum_{j=1}^{n}\left(\overline{G_{ij}} - \propto_n\right)^2\right)^{0.5}, n = 1, 2, ..., 9; \tag{50}$$

$$\underline{\updownarrow}' = \left(\sum_{j=1}^{n}\left(\underline{G_{ij}} - \propto_n'\right)^2\right)^{0.25}, n = 1, 2, ..., 9; \tag{51}$$

$$\overline{\updownarrow}' = \left(\sum_{j=1}^{n}\left(\overline{G_{ij}} - \propto_n'\right)^2\right)^{0.25}, n = 1, 2, ..., 9; \tag{52}$$

Distance between cloud's elements and two bounds of $\otimes G$ is exhibited in Fig. 4.

Step 5.4.3. The final step is the computation of GUV in accordance with Eq. (53):

$$\varrho = \left(\frac{\left(\left(\left(\underline{\updownarrow} + \overline{\updownarrow}\right) + \left(\underline{\updownarrow}' + \overline{\updownarrow}'\right)\right)\left(\left(\otimes\underline{G_{ij}} - \otimes\overline{G_{ij}}\right)^2\right)^{0.5}\right).\left(\otimes\underline{G_{ij}} - \otimes\overline{G_{ij}}\right)^2}{\left(\otimes\underline{G_{ij}}^2 + \otimes\overline{G_{ij}}^2\right)}\right)^{-1} \tag{53}$$

Hence, the S^{max} and S^{min} have been defined in Table 14.

The final section is to prioritize suppliers by Eq. (51). To compute the (V_{ij}), we have proposed in simple equation:

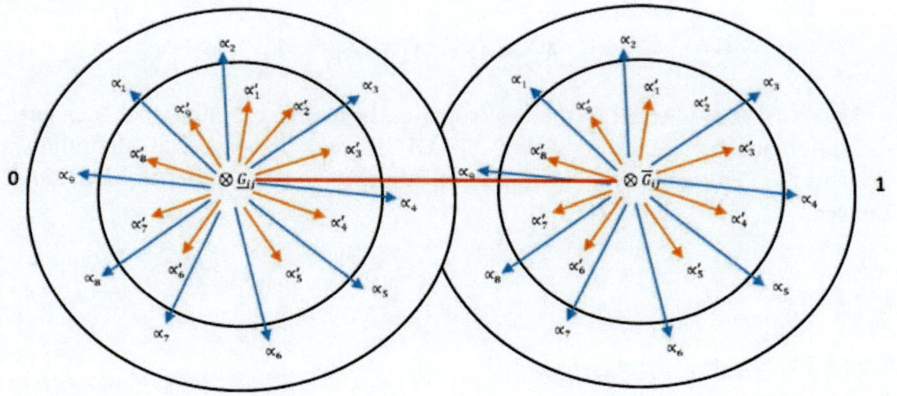

Fig. 4 The number clouds around the upper and lower bound of $\otimes G$

Table 14 The positive and negative ideal solutions

	C_1	C_2	C_3	C_4	C_5
S^{min}	[0.0464, 0.0412]	[0.1123, 0.1123]	[0.0516, 0.0512]	[0.0155, 0.0336]	[0.0436, 0.0378]
S^{max}	[0.0205, 0.0205]	[0.0179, 0.0358]	[0.1163, 0.1027]	[0.0777, 0.0674]	[0.0087, 0.0188]

Table 15 V_{ij} in accordance with Eq. (54)

	C_1	C_2	C_3	C_4	C_5
A_1	0.00252	0.005295	0.00514	0.002455	0.00407
A_2	0.00205	0.01123	0.00514	0.00536	0.003315
A_3	0.00438	0.006875	0.01095	0.00483	0.00257
A_4	0.0034	0.002685	0.008495	0.007255	0.001375

$$\mathrm{V_{ij}} = \xi.(\otimes\underline{\mathrm{G_{ij}}} + \otimes\overline{\mathrm{G_{ij}}}) \tag{54}$$

where (ξ) is the coefficient of uncertainty/probability in which in this chapter, (ξ=0.05); the results of Eq. (54) is exposed in Table 15.

With respect to Eq. (42) and Tables 14 and 15, the prioritization is:

$$\gamma_i = \begin{cases} \gamma_1 = 0.2078 \\ \gamma_2 = 0.7785 \\ \gamma_3 = 0.4135 \\ \gamma_4 = 0.3642 \end{cases}$$

then

$$A_2 > A_3 > A_4 > A_1$$

Hence (A_2) is selected as the best supplier. In the next section of this chapter, the difference between the original G-TOPSIS and the proposed algorithm is investigated. Furthermore, the impact of strategies on the supplier selection is showed.

6 Comparison

In this section, two parts of the paper are investigated. First, we implemented the original G-TOPSIS algorithm on the data and compared it with the proposed novel algorithm. According to the Grey original TOPSIS procedure (Eqs. 14–19), the following tables carry the information of each steps (Tables 16, 17 and 18).

Next step is the calculation of (S^{max}) as the ideal alternative:

$$S^{max} = \{[0.124,\ 0.124], [0.262,\ 0.262], [0.297,\ 0.297], [0.202,\ 0.202]\ [0.114,\ 0.114]\}$$

According to Eq. (18), the distance between each alternative sequence needs to be computed (Table 19).

With respect to Eq. (19), if (ξ=0.05), then, the grey relation coefficient between each of the alternatives is computed as:

Table 16 Supplier selection decision-making matrix

	C_1	C_2	C_3	C_4	C_5
A_1	[5, 6]	[3, 4]	[4, 5]	[1, 3]	[5, 6]
A_2	[4, 5]	[6, 9]	[4, 5]	[4, 5]	[4, 5]
A_3	[9, 10]	[4, 5]	[9, 10]	[3, 4]	[3, 4]
A_4	[6, 9]	[1, 3]	[6, 9]	[5, 6]	[1, 3]

Table 17 The normalized decision-making matrix

	C_1	C_2	C_3	C_4	C_5
A_1	[0.555, 0.6]	[0.5, 0.444]	[0.444, 0.5]	[0.2, 0.5]	[1]
A_2	[0.444, 0.5]	[1]	[0.444, 0.5]	[0.8, 0.833]	[0.8, 0.833]
A_3	[1]	[0.666, 0.555]	[1]	[0.6, 0.666]	[0.6, 0.666]
A_4	[0.666, 0.9]	[0.166, 0.333]	[0.666, 0.9]	[1]	[0.2, 0.5]

Table 18 The weighted normalized decision-making matrix

	C_1	C_2	C_3	C_4	C_5
A_1	[0.069, 0.074]	[0.131, 0.116]	[0.132, 0.149]	[0.040, 0.101]	[0.114, 0.114]
A_2	[0.055, 0.062]	[0.262, 0.262]	[0.132, 0.149]	[0.162, 0.169]	[0.091, 0.095]
A_3	[0.124, 0.124]	[0.175, 0.145]	[0.297, 0.297]	[0.121, 0.135]	[0.068, 0.076]
A_4	[0.082, 0.111]	[0.043, 0.087]	[0.198, 0.268]	[0.202, 0.202]	[0.023, 0.057]

Table 19 The distance between alternative sequences

	C_1	C_2	C_3	C_4	C_5	$\min_{j} \Delta_i(j)$	C_1
$d_1(j)$	0.105	0.277	0.313	0.263	0	0	0.313
$d_2(j)$	0.131	0	0.313	0.073	0.042	0	0.313
$d_3(j)$	0	0.204	0	0.148	0.084	0	0.204
$d_4(j)$	0.055	0.394	0.128	0	0.148	0	0.394
$\min_{i} \min_{j} \Delta_i(j)$						0	
$\max_{i} \max_{j} \Delta_i(j)$							0.394

$$\gamma_i = \begin{cases} \gamma_1 = 0.02015 \\ \gamma_2 = 0.0340 \\ \gamma_3 = 0.04320 \\ \gamma_4 = 0.02640 \end{cases}$$

then

$$A_3 > A_2 > A_4 > A_1$$

By comparison of two obtained results, the difference between rank of each alternatives has been illustrated in Fig. 5.

As it has been illustrated in Fig. 5, the ranks of the supplier number 1 and the supplier number 2 are equal in the two methodologies; however, the supplier number 2 possesses the first ranking in the proposed novel grey-TOPSIS method and stood in the second place in the original methodology of Grey-TOPSIS. There is a same story for supplier number 3; it possesses first rank in the proposed methodology and

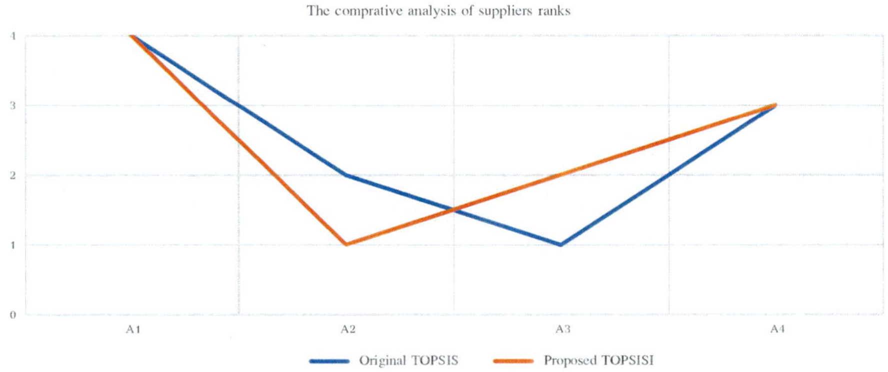

Fig. 5 The comparison between rankings of each alternatives from the proposed and original method

Table 20 The normalized supplier selection decision matrix

$W_{\otimes G_j}$	0.200932	0.196	0.20093	0.2009	0.2007
W_{DM}	[0.15, 0.15]	[0.25, 0.25]	[0.30, 0.30]	[0.175, 0.175]	[0.125, 0.125]
$\otimes W'_{\otimes G_j}$	[0.151, 0.151]	[0.245, 0.245]	[0.302, 0.302]	[0.176, 0.176]	[0.126, 0.126]
	C_1	C_2	C_3	C_4	C_5
A_1	[0.208, 0.200]	[0.214, 0.190]	[0.174, 0.172]	[0.077, 0.166]	[0.384, 0.333]
A_2	[0.166, 0.166]	[0.428, 0.428]	[0.174, 0.172]	[0.308, 0.222]	[0.308, 0.277]
A_3	[0.375, 0.333]	[0.286, 0.238]	[0.391, 0.345]	[0.231, 0.222]	[0.231, 0.222]
A_4	[0.250, 0.300]	[0.071, 0.142]	[0.261, 0.310]	[0.384, 0.333]	[0.077, 0.166]

second place in the original method, while both methodologies take the impact of (λ_j) into account.

The most important part of this section is the comparison between supplier selection with the impact of the selected strategies, which are derived from SWOT matrix by SIM method, and the selection of the alternatives without the impact of the firm's strategies. As mentioned before, in order to select the best supplier in accordance with the organization's strategies, first (λ_j) ought to be computed. In this section, we have investigated the difference between selected suppliers with the impact of (λ_j) as the value of interaction which shows the effects of the selected strategies and the evaluation of the suppliers from supplier selection procedure without the impact of the selected strategies. The supplier evaluation/selection process without taking the (λ_j) impact into account through the novel Grey-TOPSIS algorithm is given in the following tables.

With respect to Tables 20, 21, 22 and 23, the normalized decision matrix is demonstrated in Table 20.

Therefore, according to Eq. (43), the prioritization of the suppliers is as follows:

Table 21 The weighted normalized decision matrix

	C_1	C_2	C_3	C_4	C_5
A_1	[0.031, 0.030]	[0.052, 0.047]	[0.053, 0.052]	[0.014, 0.029]	[0.048, 0.042]
A_2	[0.025, 0.025]	[0.105, 0.105]	[0.053, 0.052]	[0.054, 0.039]	[0.039, 0.035]
A_3	[0.057, 0.050]	[0.070, 0.058]	[0.118, 0.104]	[0.041, 0.039]	[0.029, 0.028]
A_4	[0.038, 0.045]	[0.017, 0.035]	[0.079, 0.094]	[0.068, 0.059]	[0.010, 0.021]

Table 22 The positive and negative solutions

	C_1	C_2	C_3	C_4	C_5
S^{min}	[0.057, 0.050]	[0.105, 0.105]	[0.118, 0.104]	[0.068, 0.059]	[0.048, 0.042]
S^{max}	[0.025, 0.025]	[0.017, 0.035]	[0.053, 0.052]	[0.014, 0.029]	[0.010, 0.021]

Table 23 (V_{ij}) where ($\xi = 0.05$)

	C_1	C_2	C_3	C_4	C_5
A_1	0.0031	0.0033	0.005225	0.0021	0.0045
A_2	0.0025	0.0065	0.005225	0.0047	0.0037
A_3	0.0053	0.0044	0.011114	0.0040	0.0029
A_4	0.0042	0.0011	0.008622	0.0063	0.0015

$$\gamma_i = \begin{cases} \gamma_1 = 0.3051 \\ \gamma_2 = 0.0570 \\ \gamma_3 = 0.6847 \\ \gamma_4 = 0.4553 \end{cases}$$

then

$$A_3 > A_4 > A_1 > A_2$$

To investigate the impact of strategies on the supplier selection, the comparative analysis of the suppliers evaluation considering the impact of the selected strategies is illustrated in Fig. 6.

As illustrated in the above, there is a deep difference between rankings due to the impact of the firm's strategies. According to the results, in the process without the consideration of the strategies, supplier number 3 stood in the first place, while it possessed the second rank in the proposed method. The most alteration happened to the supplier number 2, which possesses first rank in the proposed approach affected by the firm's strategies, while stood in the last place in the process without taking the

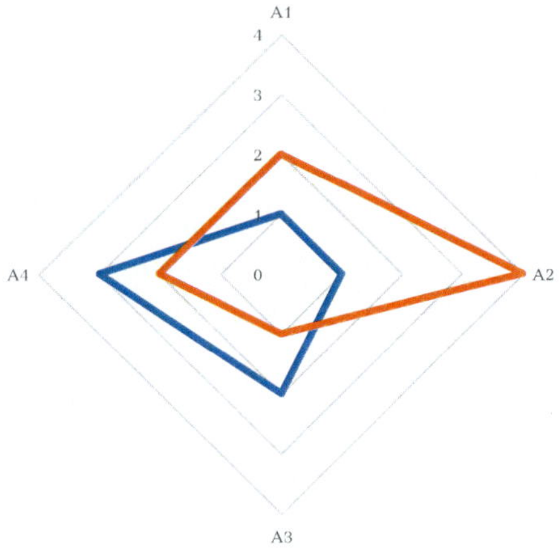

Fig. 6 Suppliers evaluation with the impact of the selected strategies (the orange lines) and without their impact (the blue lines)

firm's strategies into account. It indicates to what extent the strategies could impact the firm's internal decisions, in this case, supplier evaluation.

7 Conclusion

In this chapter, a new mathematical approach is proposed to compute the relation and interaction between two groups of irrelevant variables of decision matrices in decision-making problems using MCDM methods. To show the process of the novel approach, it is implemented on an MCDM problem, strategic supplier evaluation problem. In real-world problems, with the emergence of intensive undulations in environmental variables, decision-makers constantly encounter uncertainty. In the approach, we have benefited from the grey systems theory to deal with the uncertainty generated through the decision-making process.

In the paper, to architect the structures of irrelevant variables in decision matrices, the effects of a firm's strategies have been investigated on the evaluation and selection of the best supplier. The novel approach deals with grey form of TOPSIS and grey Entropy. To convey the effect of strategies on the suppliers evaluation, a relation matrix is used to compose the interaction between firm's strategies and the suppliers evaluation criteria. The output of the matrix used in the weighting process of those criteria in another MCDM matrix to evaluate the suppliers and select the best one. Indeed, the approach is constituted on a relation matrix between output of one matrix, in our case, the selected strategies through SWOT analysis by SIM, and criteria of another decision-making matrix which is suppliers evaluation in our case. To carry the approach, Shannon's Entropy played the main role which could potentially change for other problems. Furthermore, in this paper, we have proposed new form of Grey-TOPSIS and some transformation methods for the transforming of the grey numbers to white numbers.

In this chapter, new algorithms have been proposed, therefore, we suggest these topics for further research:

1. Application of the proposed grey entropy for objective weighting in other grey-based MCDM problems.
2. In this paper, we used many new transformation equations in each step of the proposed methodology. Researchers can develop those equations with new ideas.
3. One of the most important concepts that have been proposed in this research is the grey uncertainty value (GUV). It is a numerical platform for the comparison of grey numbers. Another exciting suggestion for future work could be the expansion of the GUV.
4. In this paper, to analyze the SWOT matrix, the grey SIM is utilized in the grey environment. To handle the uncertainty of the SWOT analysis, developing the method in fuzzy form is another interesting suggestion.

References

1. Zha, S., Guo, Y., Huang, S., Wang, S.: A hybrid MCDM method using combination weight for the selection of facility layout in the manufacturing system: a case study. Math. Prob. Eng. (2020)
2. Eyvindson, K., Öhman, K., Nordström, E.M.: Using uncertain preferential information from stakeholders to assess the acceptability of alternative forest management plans. J. Multi-Criteria Decis. Anal. **25**(1–2), 43–52 (2018)
3. Haddad, M., Sanders, D., Tewkesbury, G.: Selecting a discrete multiple criteria decision making method for Boeing to rank four global market regions. Transp. Res. Part A Policy Pract. **134**, 1–15 (2020)
4. Zhou, H., Wang, J.Q., Zhang, H.Y.: Multi-criteria decision-making approaches based on distance measures for linguistic hesitant fuzzy sets. J. Oper. Res. Soc. **69**(5), 661–675 (2018)
5. Badulescu, Y., Cheikhrouhou, N.: Evaluation of forecasting approaches using hybrid multi-criteria decision-making models. In: Proceedings of International Conference on Time Series and Forecasting (ITISE 2018) (No. CONFERENCE), 19–21 September 2018
6. Kara, S.S., Cheikhrouhou, N.: A multi criteria group decision making approach for collaborative software selection problem. J. Intell. Fuzzy Syst. **26**(1), 37–47 (2014)
7. Mediouni, A., Cheikhrouhou, N.: Expert selection for humanitarian projects development: a group decision making approach with incomplete information relations. IFAC-PapersOnLine **52**(13), 1943–1948 (2019)
8. Zavadskas, E.K., Turskis, Z., Stević, Ž, Mardani, A.: Modelling procedure for the selection of steel pipes supplier by applying fuzzy AHP method. Oper. Res. Eng. Sci. Theory Appl. **3**(2), 39–53 (2020)
9. Wang, Y., Xu, L., Solangi, Y.A.: Strategic renewable energy resources selection for Pakistan: based on SWOT-Fuzzy AHP approach. Sustain. Cities Soc. **52**, 101861 (2020)
10. Jaller, M., Otay, I.: Evaluating sustainable vehicle technologies for freight transportation using spherical fuzzy AHP and TOPSIS. In: International Conference on Intelligent and Fuzzy Systems, pp. 118–126. Springer, Cham (2020)
11. Yildirim, B.F., Mercangoz, B.A.: Evaluating the logistics performance of OECD countries by using fuzzy AHP and ARAS-G. Eurasian Econ. Rev. **10**(1), 27–45 (2020)
12. Zamani, R., Ali, A.M.A., Roozbahani, A.: Evaluation of adaptation scenarios for climate change impacts on agricultural water allocation using fuzzy MCDM methods. Water Resour. Manage. **34**(3), 1093–1110 (2020)
13. Fan, S., Zhang, J., Blanco-Davis, E., Yang, Z., Yan, X.: Maritime accident prevention strategy formulation from a human factor perspective using Bayesian Networks and TOPSIS. Ocean Eng. **210**, 107544 (2020)
14. Ocampo, L., Deiparine, C.B., Go, A.L.: Mapping strategy to best practices for sustainable food manufacturing using fuzzy DEMATEL-ANP-TOPSIS. Eng. Manage. J. 1–21 (2020)
15. Chai, J., Liu, J.N., Ngai, E.W.: Application of decision-making techniques in supplier selection: a systematic review of literature. Expert Syst. Appl. **40**(10), 3872–3885 (2013)
16. Negash, Y.T., Kartika, J., Tseng, M.L., Tan, K.: A novel approach to measure product quality in sustainable supplier selection. J. Clean. Prod. **252**, 119838 (2020)
17. el Hiri, M., En-Nadi, A., Chafi, A.: Suppliers selection in consideration of risks by a neural network. Int. J. Eng. **32**(10), 1454–1463 (2019)
18. Wu, C., Barnes, D.: A literature review of decision-making models and approaches for partner selection in agile supply chains. J. Purch. Supply Manag. **17**(4), 256–274 (2011). https://doi.org/10.1016/j.pursup.2011.09.002
19. Mosadeghi, R., Warnken, J., Tomlinson, R., Mirfenderesk, H.: Uncertainty analysis in the application of multi-criteria decision-making methods in Australian strategic environmental decisions. J. Environ. Plann. Manage. **56**(8), 1097–1124 (2013)
20. Abdel-, M., Chang, V., Gamal, A., Smarandache, F.: An integrated neutrosophic ANP and VIKOR method for achieving sustainable supplier selection: a case study in importing field. Comput. Ind. **106**, 94–110 (2019)

21. Abdel-Basset, M., Mohamed, M., Smarandache, F.: A hybrid neutrosophic group ANP-TOPSIS framework for supplier selection problems. Symmetry **10**(6), 226 (2018)
22. Zakeri, S., Keramati, M.A.: Systematic combination of fuzzy and grey numbers for supplier selection problem. Grey Syst. Theory Appl. **5**(3), 313–343 (2015)
23. Deng, J.L.: A Course on Grey System Theory. Huazhong University of Science and Technology Press, Wuhan (1990)
24. Karimi, T., Hojati, A.: Designing a medical rule model system by using rough–grey modeling. Grey Syst. Theory Appl. (2020)
25. Zhang, X., Jin, F., Liu, P.: A grey relational projection method for multi-attribute decision making based on intuitionistic trapezoidal fuzzy number. Appl. Math. Model. **37**(5), 3467–3477 (2013)
26. Deng, X., Hu, Y., Deng, Y., Mahadevan, S.: Supplier selection using AHP methodology extended by D numbers. Expert Syst. Appl. **41**(1), 156–167 (2014)
27. Li, B., Zhu, X.: Grey relational decision making model of three-parameter interval grey number based on AHP and DEA. Grey Syst. Theory Appl. (2019)
28. Laurenza, M., Consolini, G., Storini, M., Damiani, A.: A Shannon entropy approach to the temporal evolution of SEP energy spectrum. Astrophys. Space Sci. Trans. **8**(1), 19–24 (2012)
29. Zakeri, S., Delavar, M.R.R., Cheikhrouhou, N.: Dairy market selection approach using MCDM methods: a case of Iranian dairy market. Int. J. Manage. Decis. Mak. **19**(3), 267–311 (2020)
30. Zakeri, S.: Ranking based on optimal points multi-criteria decision-making method. Grey Syst. Theory Appl. (2019)
31. Lin, Y., Chen, M.Y., Liu, S.: Theory of grey systems: capturing uncertainties of grey information. Kybernetes (2004)
32. Wang, Z.X.: Correlation analysis of sequences with interval grey numbers based on the kernel and greyness degree. Kybernetes (2013)
33. Darvishi, D., Liu, S., Forrest, J.Y.L.: Grey linear programming: a survey on solving approaches and applications. Grey Syst. Theory Appl. (2020)
34. Darvishi, D., Forrest, J., Liu, S.: A comparative analysis of grey ranking approaches. Grey Syst. Theory Appl. (2019)
35. Amin, S.H., Razmi, J. and Zhang, G.: Supplier selection and order allocation based on fuzzy SWOT analysis and fuzzy linear programming. Expert Syst. Appl. **38**(1), 334–342 (2011)
36. Sanito, R.C., You, S.J., Chang, T.J., Wang, Y.F.: Economic and environmental evaluation of flux agents in the vitrification of resin waste: a SWOT analysis. J. Environ. Manage. **270**, 110910 (2020)
37. Gao, C.Y., Peng, D.H.: Consolidating SWOT analysis with nonhomogeneous uncertain preference information. Knowl.-Based Syst. **24**(6), 796–808 (2011)
38. Rauch, P.: SWOT analyses and SWOT strategy formulation for forest owner cooperations in Austria. Eur. J. Forest Res. **126**(3), 413–420 (2007)
39. Xu, D., Dong, L.: Strategic diagnosis of China's modern coal-to-chemical industry using an integrated SWOT-MCDM framework. Clean Technol. Environ. Policy **21**(3), 517–532 (2019)
40. Anser, M.K., Mohsin, M., Abbas, Q., Chaudhry, I.S.: Assessing the integration of solar power projects: SWOT-based AHP–F-TOPSIS case study of Turkey. Environ. Sci. Pollut. Res. 1–13 (2020)
41. Shahanipour, S., Amindoust, A., Sahraian, K., Beiranvand, S.: Identification and prioritization of human resource strategies with employees' creativity approach in administrative organizations using SWOT–ANP. Opsearch **57**(1), 119–143 (2020)
42. Sevkli, M., Oztekin, A., Uysal, O., Torlak, G., Turkyilmaz, A., Delen, D.: Development of a fuzzy ANP based SWOT analysis for the airline industry in Turkey. Expert Syst. Appl. **39**(1), 14–24 (2012)
43. Zakeri, S., Yang, Y., Hashemi, M.: Grey strategies interaction model. J. Strat. Manage. (2018)
44. Shannon, C.E.: A mathematical theory of communication. ACM SIGMOBILE Mobile Comput. Commun. Rev. **5**(1), 3–55 (2001)
45. Sachdeva, A., Kumar, D., Kumar, P.: Multi-factor failure mode critically analysis using TOPSIS. J. Ind. Eng. Int. Islamic Azad Univ. **5**(8), 1–9 (2009)

46. Hwang, C.L., Yoon, K.: Methods for multiple attribute decision making. In: Multiple Attribute Decision Making, pp. 58–191. Springer, Heidelberg (1981)
47. Dymova, L., Sevastjanov, P., Tikhonenko, A.: A direct interval extension of TOPSIS method. Expert Syst. Appl. **40**(12), 4841–4847 (2013). https://doi.org/10.1016/j.eswa.2013.02.022
48. Ikram, M., Sroufe, R., Zhang, Q.: Prioritizing and overcoming barriers to integrated management system (IMS) implementation using AHP and G-TOPSIS. J. Clean. Prod. **254**, 120121 (2020)
49. Vahidi, F., Torabi, S.A., Ramezankhani, M.J.: Sustainable supplier selection and order allocation under operational and disruption risks. J. Clean. Prod. **174**, 1351–1365 (2018)

Statistical Analysis of KMM Program—An Educational Intervention

Anagha Vijay Vaidya, Shilpa Bhaskar Mujumdar, Shailaja Shirwaikar, and Aradhana Kulkarni

Abstract Educational interventions are intended to help struggling students by addressing their behavioral issues and social skills. An evaluation framework has been designed to monitor and evaluate the educational intervention program implemented by the Keep Moving Movement (KMM), a Non-profit and Non-Government Organisation. The digitalization of responses is done using FormScanner by developing a software as technical support which has facilitated quick data entry with reduced errors. This paper presents the complete life cycle of the intervention process implemented by the Keep Moving Movement as a pilot study. The impact of the KMM program is analyzed using correlation analysis, factor analysis, and paired t-test. The group wise and student wise analysis of students reveals significant positive changes in positive thinking and willingness. Positive change is also observed in students' confidence, but it is not statistically significant. The findings suggest that the KMM program can be implemented to a larger set of students to improve their positive thinking and willingness, confidence qualities.

Keywords Evaluation framework · Personality · Behavior · Intervention program · Assessment

A. V. Vaidya (✉) · S. B. Mujumdar
Symbiosis Institute of Computer Studies and Research (SICSR) Symbiosis International (Deemed University) (SIU), Model Colony, Pune, Maharashtra, India
e-mail: anagha.vaidya@sicsr.ac.in

S. B. Mujumdar
e-mail: shilpa.mujumdar@sicsr.ac.in

S. Shirwaikar
Department of Mathematics Savitribai, Phule Pune University, Pune, India

A. Kulkarni
Consultant Chezuba, Corporate Social Responsibility Consultant Pune, Pune, India
e-mail: aradhana@chezuba.net

A. J. Kulkarni (ed.), *Multiple Criteria Decision Making*, Studies in Systems, Decision and Control 407, https://doi.org/10.1007/978-981-16-7414-3_11

1 Introduction

Educational interventions are used to help students struggling with learning or emotional problems. An instructional intervention designed by a Non-profit and Non-Government Organisation (NGO), namely Keep Moving Movement (KMM), is intended to imbibe good values among the targeted students. In this study, the intervention program is referred to as the 'KMM intervention Program'.

The main objectives of monitoring and evaluation of any program or intervention are

1. To check whether the intervention works and has the right impact.
2. To refine the delivery of the program to further improve its impact.
3. To provide evidence for continuing support for the program.
4. To check the appropriateness of the program to the targeted population.
5. To identify concerns related to its implementation.

The evaluation framework must be constructed and executed along with the designing and implementation of the intervention. Technology should be used as and when possible to reduce the efforts toward data collection, data entry, and to facilitate performance evaluation.

The evaluation may use different techniques. The appropriateness of the technique depends on the objectives of the intervention to be evaluated. The technique itself can be evaluated using a pilot study. The evaluation may be on different perspectives, depending on intervention objectives. The most common evaluation perspectives are impact assessment and outcome measurement.

Impact assessment is the assessment of intervention program on students. The impact assessment is done by measuring the changes in the target population by administering a questionnaire before and after intervention. The same instrument should be used to avoid any bias. The before-after study instead of a controlled-designed study is often the simplest and cheapest method of assessing impact.

The program outcome is measured by selecting proper latent variables and the corresponding measurement indicators. The attributes to be measured can be identified through the objectives of the educational intervention. Then the instrument can be designed to measure the intended attributes.

The following five phases are defined for evaluation process of intervention program.

1. Identify the measurement indicators from intervention objectives.
2. Create an assessment questionnaire.
3. Validate the questionnaire using a pilot study.
4. Modify the questionnaire and measurement indicators according to the pilot study.
5. Assess and analyze the targeted population using the modified questionnaire.

This paper presents the administration of Educational intervention. The next section presents the background and related work. The steps of the evaluation process are described in Sect. 3. The validation step is further detailed and demonstrated using experimental data in Sect. 4. The impact and outcome assessment techniques and analytics are presented in Sect. 5. The paper ends with a conclusion and references.

2 Background and Related Work

Education mainly comprises different activities that prepare students toward academic excellence. It also has a larger goal of character building and personality development so that learners grow into responsible and constructive members of society. Personality development is not a by-product but requires time and diligent efforts. The independent personality is developed by academic excellence and other characteristics like promptness, flexible attitude, preparedness to learn, responsiveness, enthusiasm, etc. Adoption of these characteristics make a person accept responsibility, be a good listener, practice humanity, be enthusiastic about life, etc. [13]. The education imparting these skills assists the individual to become a good decision-maker, build leadership skills, achieving goals, and a positive attitude toward life [17]. It is difficult to develop personality skills with academic excellence because of time and other constraints. This necessitates the need for additional programs as educational interventions with complete focus on imparting the above skills [24].

2.1 Educational Intervention

The Educational intervention program focuses on interpersonal skills, practical skills, academic, cognitive, behavioral, and social skills which directly affect the student's competence to accomplish education [20]. The essential skills for people of the twenty-first century, as defined by cambridgeinternational.org [8], are ways of thinking, ways of working, tools for working, and skills for living in the world. Creativity, innovation, critical thinking, problem-solving, and decision-making skills are required for ways of thinking. Communication and interpersonal skills are important for ways of working. Further behavioral and social skills are essential for living in the world (cambridgeinternational.org [8]—Chapter 1–2017). The above-mentioned skills are also required for academic achievement [20]. Previous research had recorded that students with deprived skills tend to perform poorly in school [20]. Hence, academic experts have given importance to school-based interventions that focus more on functionality skills that will improve the students' achievement [26]. Most research has evidence of the association between skill development programs and achievements in academics [2, 10]. The various intervention programs in higher education for health science, nursing, and teaching medical emergencies [23] have been implemented and documented. However, intervention studies for school children have been rare.

2.2 Intervention Evaluation

There is no straightforward instrument to measure the latent variables such as skill and attitude. These are measured through the questionnaire that consists of questions related to self-view and others' views of behavioral change [2]. The self-view of behavior implies asking people to evaluate their behavior. Honest evaluation for oneself is a challenge. The others view method overcomes the drawback of self-view. In this method, peers report behavioral changes observed amongst them [22].

The broad term personal development further includes self-fulfilment, developing self-awareness, finding an identity, discovering a passion or improving the quality of life. Realizing dreams or fulfilling ambitions is all part of personal development [4]. This category also includes professional development such as the development of work-related skills. Professional self-development has grown in popularity and has helped many managers and executives to obtain better qualified and motivated staff [7].

In literature, different models have been proposed but most prominently emphasized ones are Big Five-Factor Model [9], HEXACO model, and NEO Personality Inventory [29].

Big Five personality framework is an appropriate framework through which social and emotional skills can be organized. The Big Five factors include conscientiousness (work ethic; organization), agreeableness (kindness; empathy), emotional stability (composure; flexibility), openness (curiosity; analytical thinking), and extraversion (sociability; assertiveness) [9, 29]. Another model is NEO Personality Inventory, which was originally created as a 3 factor model assessing Neuroticism, Extraversion, and Openness. The HEXACO model of personality structure summarizes human personality characteristics in terms of six dimensions, or factors: Honesty-Humility (H), Emotionality (E), Extraversion (X), Agreeableness (A), Conscientiousness (C), and Openness to Experience (O). For this measurement Zuckerman-Kuhlman Personality questionnaire is used [3, 11].

The five characteristics of a good questionnaire are simplicity, reliability, content validity, context validity, and sensitivity to change. Out of these, simplicity and reliability are important characteristics [12]. The simplicity stresses on the clarity of the questions and ease of interpretation. Reliability measures the proportion in the variation of measurements which occurs due to the diversity of values [12]. There are different statistical techniques for examining the questionnaire validity and reliability [1, 18].

But proper evaluation techniques are rarely discussed for personality development programs Xiong P et al. [30].

The present paper proposes a framework for an intervention program to improve the above-mentioned student skills. The study presents the life cycle of an intervention program from design, evaluation up to implementation. A pilot study of KMM intervention is used to demonstrate the complete process.

3 Methodology

The proposed research goals are

1. To develop a framework to execute and evaluate an intervention program
2. To observe changes in personality due to intervention program.

The current study proposes an evaluation framework to help the KMM NGO to execute and evaluate an intervention program. A questionnaire is designed to evaluate the improvement in personality skills. The code is written for OMR sheet reader using software Formscanner and collected data was digitalized using the same. The effect of the program is examined through paired t-test.

3.1 Method

KMM is connected with corporation (or ZP) schools, private schools, and junior colleges mainly catering to students from low income group families of nine states. The children to these schools come from deprived/below average income families. KMM works to help them to improve their personality skills. As a part of the personality development program KMM conducts seven sessions within six months for children of 9th and 10th standard. The two most important reasons for failure and lack of self-belief are "Inadequate Preparations" and "Negative Attitudes" (IPNA). While the students are huge bundles of potential, IPNA in them is really alarming. KMM via volunteers is trying to overcome this issue for the last 20 years. Innovative approaches are used to help create an impact on the students.

The session starts with a motivational story and different activities are conducted to know their moral values and learning. Activities like enacting a play, debates, discussions are conducted followed by discussion regarding the values communicated through these activities.

The facilitators of the KMM program then discuss the consequences of various actions that can be taken in that situation and reach a message telling the right action. The procedure to build the evaluation process is discussed in Sect. 3.2.

3.2 Evaluation Framework

The evaluation framework consists of tasks from designing of questionnaire up to analysis of data and is a multi-step process as presented in Fig. 1.

The steps are described below:

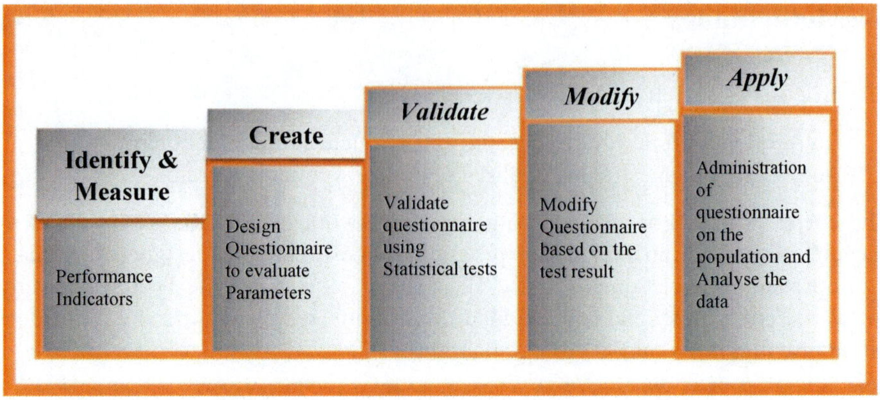

Fig. 1 Multistep process for evaluation framework

Step 1: Identify and Measure the Performance Indicators

The program outcome is measured by selecting proper performance Indicators for the latent attributes of the targeted population that are addressed by the KMM program. This program conveys right values through storytelling, quotes, proverbs, role play, role models, etc. The essence of the program is to imbibe basic essential qualities for achieving academic excellence [13]. The qualities that formulate the KMM objective and associated measurement parameters for these latent variables are given in Table 1. The measurement parameters are decided by the KMM organization.

Step 2: Create and Evaluate the Instrument

Once measurement indicators are finalized, the next step is to design the questionnaire and evaluate it. Following two steps are followed for designing the questionnaire.

1. Content Identification
2. Digitalization of questionnaire.

Content Identification: The questionnaire is designed by considering the above nine performance indicators. It is always a challenge to measure behavioral characteristics. Therefore, the questions were constructed considering students' reactions relating to nine performance indicators for the same problem at home, school, and social environment. The questionnaire wording is kept simple to convey the intended meaning. The Likert scale (1–3) is used to design the questionnaire. The questionnaire is attached in Appendix-I.

Digitalization of Questionnaire: Data collection is done by using Optical Mark Recognition (OMR) sheets and digitalized by FormScanner software (open-source Version 1.1.3. (FormScanner (n.d.)). The use of OMR sheet and FormScanner has facilitated the fast conversion of data into digital form which reduces delay in data analysis.

Table 1 Performance parameters and their measures

S. No.	Performance indicator	Purpose	Measurement parameters
1	Confidence	It encourages accepting of challenges in life, gives freedom from anxiety, fear, and feels self-worth. It creates motivation for task achievement	Self-awareness, Decisiveness, Friendly, Acceptance, aware of self-abilities
2	Consistency	It helps to achieve the goal of life or task. It identifies the effectiveness of the efforts towards the goal	Doing the task effectively at any time in any situation, dedicated to the objective, focused on my work
3	Positive thinking	A positive attitude is a guide to leading a positive life. It develops creative, constructive thinking, and motivation to do things and accomplish goals	Positive approach towards the situation, Motivating friends, and self in any critical situation
4	Leadership	It teaches effective communication skills, the ability to solve the problems, develop a wide vision in life, and develop decision-making ability	Taking initiative, lead any activity with confidence and self-awareness, provide opinions and judgment about any situation
5	Self-Motivation	It drives to accomplish tasks and goals. It sets priorities in life, teaches perseverance, to fight against fear and build self-confidence	Doesn't bow to peer pressure, Confident in self-judgment, Prepared to stick with own opinions
6	Good habits	Good habits will bring positivity in life. Help to stay focused on the goal	Focused on own work, Not easily distracted
7	Discipline	Discipline brings stability and structure into a person's life	Self-control, managing time in the most effective way, ability to focus on objectives
8	Willingness	Desire, wish or readiness to acquire new knowledge leading to self- development for converting dreams to reality	Accept the challenges and willing to move into untried areas, Confidence in their own opinions, and judgment
9	Goal-Oriented	To stay in focus and motivated to achieve goal	To identify idol

The OMR sheet is designed such that bubbles and answers are displayed on the same sheet to avoid student's confusion while marking the answers as shown in Fig. 2. The FormScanner scans the sheet properly if it is kept straight, and bubbles are to be made dark black colored. Any deviation in the position of answer paper prevents scanning. Hence, the software was updated with codes so that the answer paper can be scanned irrespective of deviation in the position of the answer paper.

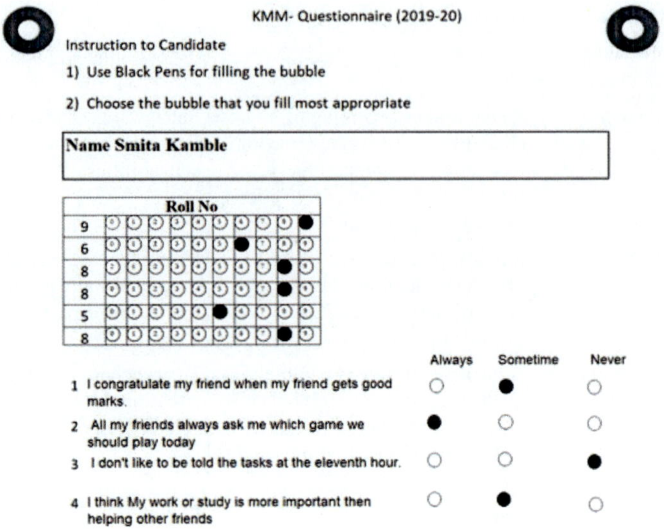
KMM- Questionnaire (2019-20)

Instruction to Candidate

1) Use Black Pens for filling the bubble

2) Choose the bubble that you fill most appropriate

Name Smita Kamble

Roll No

9
6
8
8
5
8

		Always	Sometime	Never
1	I congratulate my friend when my friend gets good marks.		●	
2	All my friends always ask me which game we should play today	●		
3	I don't like to be told the tasks at the eleventh hour.			●
4	I think My work or study is more important then helping other friends		●	

Fig. 2 Sample of questionnaire design

Step 3: Validating Evaluation Sheet

The questionnaire is validated through Content analysis (feasibility of questionnaire), Construct Validity, and Reliability (stability and internal consistency) (Internal consistency. (n.d.)). Content analysis is measured by the subject experts. The construct validity and reliability are measured through different statistical techniques. The evaluation of the questionnaire is done using the Cronbach alpha test. Construct validation is performed by Principal Component Analysis (PCA). The reliability of the questionnaire is measured through the component or factor loading and Cronbach alpha test [6, 12, 25]. The validation process is discussed in detail in Sect. 4.

Step 4: Modify the Questionnaire

Based on Cronbach alpha results, the questionnaire is modified to drop the redundant questions and continue the relevant questions.

Step 5: Administrate the questionnaire and analyse data

It is a two-step process, data administration, and analysis.

(a) **Administration of the questionnaire**: The process consisting of designing questionnaire till data analysis is depicted in Fig. 3. The questionnaire is administered on the target population. The questionnaire data is collected before and after the KMM program implemented. The process as explained in Fig. 3 makes use of available technology. The filled forms are scanned using FormScanner with 300 DPI and black and white colors only. It is a feeder scan and makes sure all sheets are properly placed straight. The data read is converted into CSV file, and required data pre-processing is performed.

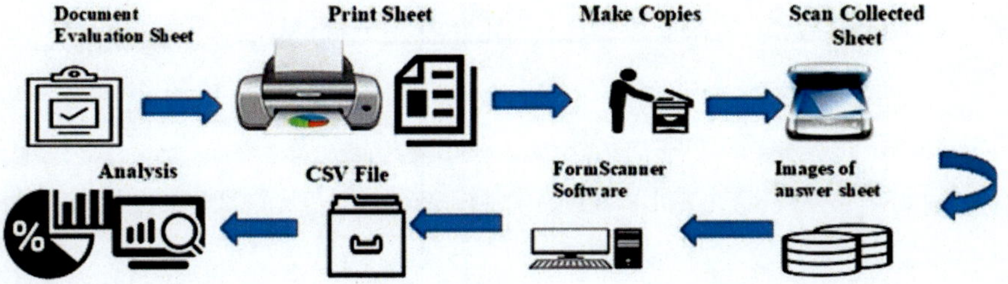

Fig. 3 Administrative process

(b) **Data Analysis**: The descriptive, predictive, and prescriptive analysis is performed on collected data. The details are described in Sect. 5.

3.3 Questionnaire Validation

The questionnaire designed to test the attitudes of school children consists of 40 questions (Appendix 1). Summary of number of questions corresponding to each skill is presented in the Table 2.

The questionnaire was validated by using Cronbach's Alpha test [16]. The modified questionnaire is used for data collection and digitalization of data is done using FormScanner.

4 Analysis, Result, and Discussion

The present section discusses the validation results of questionnaires using Cronbach alpha test followed by descriptive statistics for categorical variables, correlation analysis, factor analysis, and impact analysis (using z-test and paired t-test).

4.1 Cronbach Alpha Test

The Cronbach alpha test is carried out by using "psyc" package in R-programming before and after the improvisation in questionnaire. The results of test are presented in Table 3.

The alpha value 0.59 indicates poor reliability as per the standards specification. (Internal consistency. (n.d.); [28]). The seven questions were dropped from the questionnaire on the basis of Cronbach alpha test. The remaining 33 questions of the set were again given to collect the responses from the students. The observed Cronbach

Table 2 Question summary of skill parameters

Parameters	X1	X2	X3	X4	X5	X6	X7	X8	X9
Skill	Confidence	Consistency	Positive thinking	Leadership	Self motivation	Good habits	Discipline	Willingness	Goal
Total questions	7	4	6	8	3	2	4	3	3

Table 3 Reliability analysis

Parameters ↓	Before	After data cleaning
Sample size	43	43
Number of items(questions)	40	33
Cronbach's Alpha	0.595	0.735

alpha value for revised questionnaire is 0.735 which is an acceptable range. Now skill scores are analyzed using descriptive statistics.

4.2 *Descriptive Statistics to Study Impact of KMM Program*

The descriptive statistics of skill scores before and after KMM is represented in Tables 4 and 5 respectively.

The skewness and kurtosis scores (skewness <2, kurtosis <7) from Tables 4 and 5 show data follows normal distribution [21, 27].

The distribution of the above data is visualized through violin plots (Figs. 4 and 5) before and after the intervention of the KMM program.

The violin plot indicates some changes in data value after the intervention of the program. The data change has been observed especially for the attributes Consistency, Positive Thinking, Good Habits, Discipline, and Willingness. As the flatter end on the upper side of the plot indicates most of the students have scored higher marks. This indicates the activities of KMM program pertaining to these skills are effective.

No change is observed in the distribution pattern of 'Confidence', 'Leadership', 'Self-Motivation', and 'Goal-Setting '. Probably these attributes are inherited from childhood and may require additional efforts to change them.

Table 4 Pre-program scores

S. No.	Skill	Min	Q1	Median	Mean	Q3	Max	SD	Skewness	Kurtosis
1	Confidence	0.29	0.5	0.5	0.56	0.64	0.86	0.12	0.34	−0.19
2	Consistency	0.35	0.6	0.7	0.69	0.75	1	0.15	−0.36	−0.22
3	Positive thinking	0.35	0.69	0.75	0.74	0.81	1	0.12	−0.52	1.04
4	Leadership	0.18	0.43	0.5	0.51	0.59	0.77	0.14	−0.38	−0.67
5	Self motivation	0	0.42	0.5	0.57	0.67	1	0.23	−0.09	−0.32
6	Good habits	0.44	0.63	0.75	0.74	0.85	1	0.15	−0.43	−0.68
7	Discipline	0.25	0.58	0.7	0.7	0.83	1	0.18	−0.46	−0.34
8	Willingness	0.38	0.7	0.8	0.77	0.9	1	0.18	−0.36	−0.74
9	Goal	0.17	0.33	0.67	0.6	0.75	1	0.24	0.25	−1.04

Table 5 Post programming scores

S. No.	Skill	Min	Q1	Median	Mean	Q3	Max	SD	Skewness	Kurtosis
1	Confidences	0.79	1	1.07	1.07	1.14	1.43	0.16	0.2	2.95
2	Consistency	0.67	1	1.25	1.86	1.38	1.5	0.23	−0.39	2.28
3	Positive thinking	1.08	1.3	1.33	1.32	1.42	1.5	0.13	−0.23	2.22
4	Leadership	0.71	0.9	1	1	1.09	1.25	0.14	−0.25	2.22
5	Self motivation	0.5	0.9	1	1.06	1.17	1.5	0.22	−0.08	2.49
6	Good habits	0.75	1	1.25	1.17	1.25	1.5	0.22	0.01	2.95
7	Discipline	0.75	1	1.25	1.17	1.33	1.5	0.21	−0.68	2.15
8	Willingness	1	1.3	1.5	1.4	1.5	1.5	0.13	−1.06	2.28
9	Goals	0.5	0.8	1.17	1.06	1.17	1.5	0.25	0.05	3.69

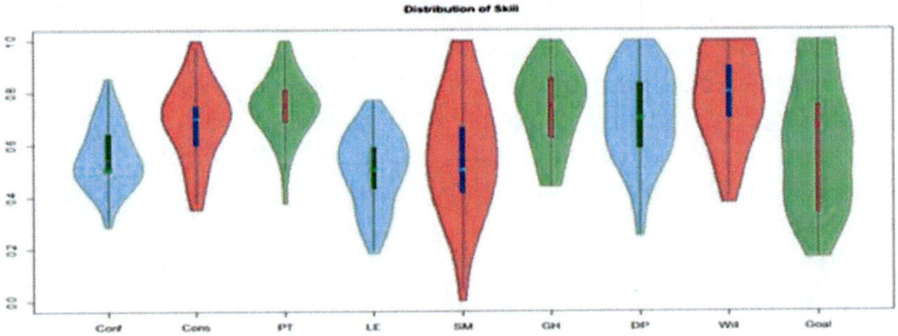

Fig. 4 Violin plot—before the intervention of KMM program

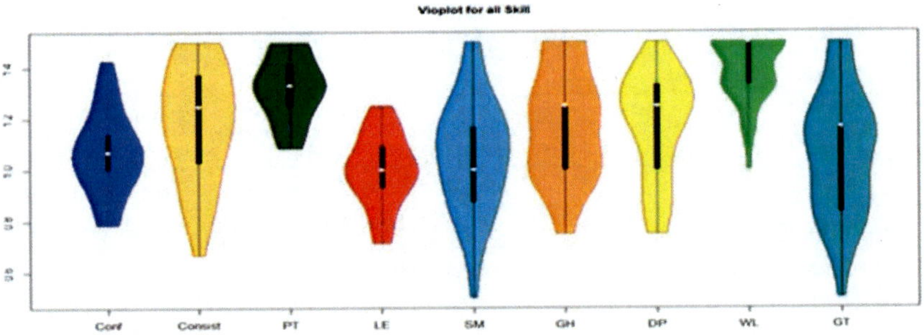

Fig. 5 Violin plot—after the intervention of KMM program

4.3 Correlational Analysis

To study the association between various skill parameters under consideration, correlations are calculated among them. The correlations between the attributes and corresponding p-values are presented in Table 6.

The correlation matrix shows that the Confidence is significantly related to leadership, self-motivation, and good habits; Consistency is significantly related to leadership, self-motivation, discipline, willingness; Positive thinking is significantly related to willingness; Leadership is significantly related to self-confidence, good habits, discipline; Self-motivation is significantly related to discipline and willingness; Good habits are significantly related to discipline and willingness; Discipline is significantly related to willingness.

In order to identify the factors underlying, the mutual association between skill parameters factor analysis is performed.

4.4 Factor Analysis

Exploratory Factor analysis (EFA) is carried out to identify latent relation structure among the set of skill parameter variables hence to reduce the number of variables (Preetish, (n.d.)). The factor map analysis identifies the quality of variables to serve the objective of the survey. The factor map diagram shows most related variables in the same colors and further descending orders of colors in the legend represents the degree of contribution of variables in EFA. The factor map diagram is depicted in Fig. 6.

Figure 7 shows contributions of variables X9, X1, (X6, X8), (X7, X2), and (X3, X5, X4) in ascending order. The variables indicated in red color (X3, X5, X4) i.e. (Positive thinking, Leadership, and Self-motivation) show the highest contribution. Since variable X9 (Goal) has the least contribution; hence, it is omitted for further analysis.

The extracted variables and their quality of groups depending on their distances from the circumference are represented in Fig. 7.

The four factors extracted from EFA as shown in Fig. 7 are Self-Management (F1), Positivity (F2), Good-leadership (F3), and Confidence (F4). Self-Management describes motivation, consistency, and discipline; Positivity describes positive thinking and willingness; Good-leadership describes good habits and leadership. Confidence appears as a separate factor.

The first part of Fig. 7 indicates the quality of these factors. Hence the quality of factors F1 (Self-Management), F2 and F3 (i.e. Positivity and Good leadership), and F4 (Confidence) is very high, good and poor.

The correlation matrix between significant factors is presented in Table 7.

The factors with p-value less than 0.05 are significant factors. Now the impact of the KMM program is analyzed using z-test for proportion and paired t-test.

Table 6 Correlation analysis

	Confidence	Consistency	Positive thinking	Leadership	Self motivation	Good habit	Discipline	Willing	Goal
	X1	X2	X3	X4	X5	X6	X7	X8	X9
X1	1	0.27 **(0.08)	0.06 (0.69)	0.37 *(0.01)	0.55 * (0.00)	0.34 *(0.02)	0.27 **(0.07)	0.21 (0.17)	0 (0.99)
X2		1	0.24 (0.12)	0.63 *(0.00)	0.65 * (0.00)	0.68 * (0.00)	0.83 *(0.00)	0.36 *(0.02)	0.09 (0.57)
X3			1	0.19 (0.22)	0.23 (0.14)	0.29 ** (0.06)	0.14 (0.37)	0.47 *(0.00)	0.16 (0.29)
X4				1	0.48 *(0.00)	0.41 *(0.01)	0.58 *(0.00)	0.27 (0.08)	0.13 (0.42)
X5					1	0.43 * (0.00)	0.48 *(0.00)	0.17 (0.27)	0.03 (0.85)
X6							0.83 *(0.00)	0.32 *(0.04)	0.08 (0.62)
X7							1	0.38 *(0.01)	0.04 (0.80)
X8								1	0.13 (0.83)
X9									1

*Indicates correlation significant at 5% and **Indicates correlation significant at 10%

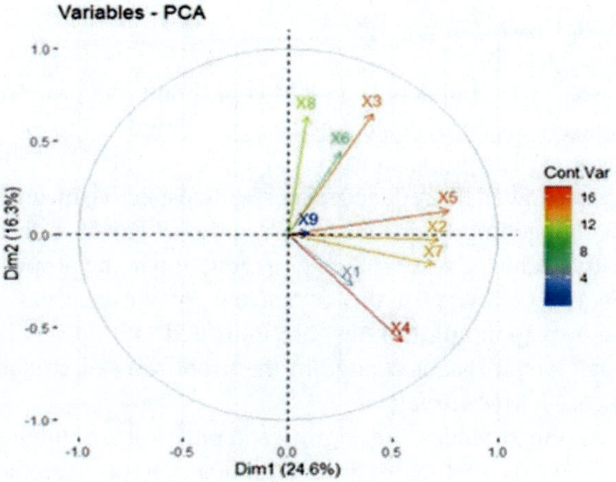

Fig. 6 Factor map diagram

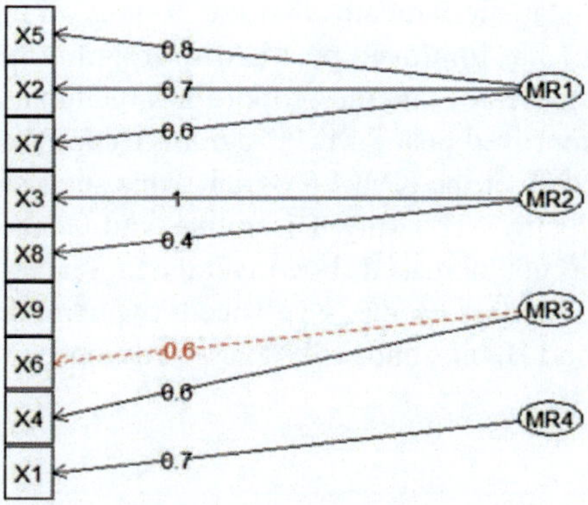

Factors	**Parameters**	**Skill**
Self-Management (F1)	X5	Self-Motivation
	X2	Consistency
	X7	Discipline
Positivity (F2)	X3	Positive thinking
	X8	Willingness
Good-leadership (F3)	X6	Good Habits
	X4	Leadership
Confidence (F4)	X1	Confidence

Fig. 7 Exploratory factor analysis

Table 7 Correlation between factors

Factor	Correlation	p.value
F3, F2	0.8214	0.00*
F1	0.7376	0.00*
F4	0.4834	0.00*

*Indicates significance at 5% level of significance

4.5 *Impact Analysis of KMM Program*

The descriptive statistics for skill score is presented to judge the normality. Impact of KMM program is studied through z-test for proportions and paired t-test.

4.5.1 Impact Analysis Using z-Test

To analyze the group wise impact of the KMM program, the proportion of students scoring less than average score is calculated and presented in Table 8 for pre (p1) and post (p2) KMM scores. The proportions are calculated for all nine skills. The students' post-KMM percent showed the reduced scores with respective mean. Further to identify a significant change in proportions for pre and post KMM, z-test of proportions is used. The positive value of z-test indicates a reduction in the proportion of students scoring less than average in a post-survey hence defined as success.

The p-values corresponding to z-test are indicated in Table 8. The p-value less than 0.10 indicates a significant change in the proportion of students scoring less than average in post-KMM survey.

Table 8 shows the p-values are significant only for confidence and positive thinking. This indicates that there is a significant improvement for these two attributes. The p1 and p2 being cumulative proportion of students scoring less than average score; the larger value of p1 will result in a positive value of Z, which shows decrease in proportion of students scoring less than average score in post KMM scores. Hence a positive value of Z-statistic indicates that the proportion of students with less than average performance has improved post KMM program. In other words, a positive value of Z-statistic indicates that the proportion of students with more than average performance has improved post KMM program. Hence the positive value of Z is considered as the success of the KMM program. This success is observed for Confidence, Consistency, Positive Thinking, Discipline, and Goals. But the proportion of students post KMM has not increased above average for leadership, self-Motivation, Good Habits, and Willingness. Hence, NGO needs to put more effort into Leadership, Self-Motivation, Good Habits, and willingness activities.

Table 8 Groupwise student score analysis

S. No.	Skill	p1	p2	Z -value	p-Value	Result
1	Confidence	0.6279	0.3902	3.00	0.05**	Success
2	Consistency	0.6279	0.4878	1.78	0.54	Success
3	Positive thinking	0.6047	0.3659	3.02	0.05**	Success
4	Leadership	0.4884	0.6341	−1.86	0.53	Failure
5	Self- motivation	0.3256	0.5366	−2.69	0.50	Failure
6	Good habits	0.4186	0.4634	−0.57	0.78	Failure
7	Discipline	0.5349	0.3902	1.84	0.53	Success
8	Willingness	0.4419	0.4878	−0.58	0.78	Failure
9	Goal	0.4651	0.4634	0.02	0.99	Success

**Indicates significant values at 10%; p1 = proportion of students scoring less than average score before KMM and p2 = proportion of students scoring less than average score post KMM

Table 9 Student wise paired t-test results

S. No.	Skill	p-value
1	Confidence	0.753
2	Consistency	0.274
3	Positive thinking	0.006*
4	Leadership	0.322
5	Self-motivation	0.107
6	Good habits	0.287
7	Discipline	0.531
8	Willingness	0.008*

*Indicates significant skill at 5% level of significance

Z-test for proportion shows a significant change in confidence and positive thinking with p-value 0.05. Though p-values are not significant for attributes consistency, discipline, and Goal, the positive value of z indicates program is effective. But leadership, self-motivation, good habits, and willingness attributes have not change post KMM program. Hence the KMM program has helped in improving five skills out nine under consideration. The next section analyses student wise KMM program impact.

4.5.2 Impact Analysis Using Paired t-Test

The paired t-test is used to test the hypotheses regarding improvement in Confidence, Consistency, Positive thinking, Leadership, Self-motivation, Good habits, and Discipline in students. The paired t-test is applied to pre and post KMM scores of 54 students. This helps to identify the skill wise influence of KMM intervention program. The paired t-test results are displayed in Table 9.

The p-value of positive thinking and Willingness being less than 0.05, we conclude that the KMM intervention program reflects the changes in Positive thinking and Willingness. The results analyzed student wise show that the program has significantly affected Positive Thinking and Willingness.

Though the group wise analysis shows no impact on Willingness.

4.6 Discussion

An educational intervention, through the KMM program, is implemented to help struggling students by addressing behavioral issues and social skills. Further it is monitored and evaluated for its efficacy. The use of FormScanner software to digitize data has reduced data entry time and errors. The measurement questions for the defined parameters are validated using the Cronbach alpha test. Correlation analysis revealed association between the nine skill parameters under study. Factor analysis has reduced nine skill parameters to three factors namely Self-Management, Positivity, and Leadership. The group wise analysis of students shows that a positive change in confidence and positive thinking has been achieved, through the KMM program. Student wise analysis shows significant changes in positive thinking and willingness, which can be attributed to the KMM program.

The successful implementation of KMM meets the first research goal of proposing a framework to execute and evaluate an intervention program. The z-test and paired t-test results to observe changes in personality due to intervention program addresses the second research goal.

4.7 Practical Implication

The framework proposed in this paper can help any NGO to execute and evaluate an intervention program. Further, the process of digitalization of data collected through the intervention program can be easily uploaded in spreadsheets which will facilitate data analysis. This overcomes the problems faced by NGOs regarding data entry and henceforth data analysis tasks.

The digitalization method of data is cost effective, consumes less man hours hence very useful for the NGOs. The KMM framework will help the development sector, grass root NGOs to bring in the authentic data on the table thus improving the overall impact, accountability, and performance of the sector.

5 Conclusion and Future Scope

The findings suggest that the present KMM program improvises positive thinking student wise and among the group of students. The improvement in confidence and willingness is seen group wise and student wise respectively. KMM program needs enhancement for improvement in leadership, self-motivation, and good habits.

This program can be implemented to a larger set of students to improve the Positive thinking, Confidence, and Willingness.

References

1. Acock, A.C.: Discovering structural equation modeling using Stata. College Station, TX: Stata Press.O'Rourke, Norm, and Larry Hatcher. (2013). A step-by-step approach to using the SAS for factor analysis and structural equation modeling, 2nd edn. SAS Institute, Cary, NC (2013)
2. Alloway, T.P., Banner, G.E., Smith, P.: Working memory and cognitive styles in adolescents' attainment. Br. J. Educ. Psychol. **80**, 567–581 (2010). https://doi.org/10.1348/000709910X494566

3. Aluja, A., Rossier, J., García, L.F., Angleitner, A., Kuhlman, M., Zuckerman, M.: A cros-cultural shortened form of the ZKPQ (ZKPQ-50-c) adapted to English, French, German, and Spanish languages. Personality Individ. Differ. **41**, 619–628 (2006)
4. Antoñanzas, J.L.: The relationship of personality, emotional intelligence, and aggressiveness in students: a study using the big five personality questionnaire for children and adults (BFQ-NA). Eur. J. Invest. Health Psychol. Educ. **11**(1), 1–11 (2021)
5. Ashton, M.C., Lee, K.: Objections to the HEXACO model of personality structure—and why those objections fail. Eur. J. Pers. **34**(4), 492–510 (2020)
6. Boparai, J.K., Singh, S., Kathuria, P.: How to design and validate a questionnaire: a guide. Curr. Clin. Pharmacol. **13**(4), 210–215 (2018)
7. Bore, M., Laurens, K.R., Hobbs, M.J., Green, M.J., Tzoumakis, S., Harris, F., Carr, V.J.: Item response theory analysis of the big five questionnaire for children-short form (BFC-SF): a self-report measure of personality in children aged 11–12 years. J. Pers. Disord. **34**(1), 40–63 (2020)
8. Cambridgeinternational.org: Developing the Cambridge learner attributes (2015). https://www.cambridgeinternational.org/support-and-training-for-schools/teaching-cambridge-at-your-school/cambridge-learner-attributes/
9. Caprara, G.V., Barbaranelli, C., Borgogni, L., Perugini, M.: The "big five questionnaire:" a new questionnaire to assess the five factor model. Personality Individ. Differ. **15**(3), 281–288 (1993). https://doi.org/10.1016/0191-8869(93)90218-R
10. Clark, C.A.C., Pritchard, V.E., Woodward, L.J.: Preschool executive functioning abilities predict early mathematics achievement. Dev. Psychol. **46**, 1176–1191 (2010). https://doi.org/10.1037/a0019672
11. Conover, A.: Zuckerman-Kuhlman Personality Questionnaire (ZKPQ). Wiley Encycl. Pers. Individ. Differ. Meas. Assess. 351–355 (2020)
12. de Yébenes Prous, M.J.G., Salvanés, F.R., Ortells, L.C.: Validation of questionnaires. Reumatología Clínica (English Edition) **5**(4), 171–177 (2009)
13. Dumrongsuntithum, C., Wongleedee, K.: Benefits of personality development training. In: Proceeding of the ICBTS. (2019)
14. Preetish (n.d.) Exploratory Factor Analysis in R. https://www.promptcloud.com/blog/exploratory-factor-analysis-in-r/
15. FormScanner (n.d.) Referred from http://www.formscanner.org/
16. Gliem, J. A., Gliem, R. R.: Calculating, interpreting, and reporting Cronbach's alpha reliability coefficient for Likert-type scales. In: Midwest Research-to-Practice Conference in Adult, Continuing, and Community Education. (2003)
17. Gündoğdu, Y.B., Turan, Y.: Evaluation of critical periods during the development of the personality In terms of religious education. ODÜ Sosyal Bilimler Araştırmaları Dergisi (ODÜSOBİAD) **8**(1), 229–239 (2018)
18. Harris, R.C., Rosenberg, L., O'Rourke, M.E.G.: Addressing the challenges of nursing student attrition. J. Nurs. Educ. **53**(1), 31–37 (2013)
19. Internal consistency (n.d.). https://en.wikipedia.org/wiki/Internal_consistency
20. Jacob, R., Parkinson, J.: The potential for school-based interventions that target executive function to improve academic achievement: a review. Rev. Educ. Res. **85**(4), 512–552 (2015)
21. Kim, H.Y.: Statistical notes for clinical researchers: assessing normal distribution (2) using skewness and kurtosis. Restorative dentistry endodontics **38**(1), 52–54 (2013)
22. Lovelace, M., Brickman, P.: Best practices for measuring students' attitudes toward learning science. CBE—Life Sci. Educ. **12**(4), 606–617 (2013)
23. Schweizer, M.L., Braun, B.I., Milstone, A.M.: Research methods in healthcare epidemiology and antimicrobial stewardship—quasi-experimental designs. Infect. Control Hosp. Epidemiol. **37**(10), 1135–1140 (2016)
24. Sharma, B., Patidar, J.: Effects of different school environment on personality development of rural girls students in Ratlam District, India. Int. J. Curr. Microbiol. App. Sci **7**(2), 411–416 (2018)

25. Tsang, S., Royse, C.F., Terkawi, A.S.: Guidelines for developing, translating, and validating a questionnaire in perioperative and pain medicine. Saudi J. Anaesth. **11**(Suppl 1), S80 (2017)
26. Ursache, A., Blair, C., Raver, C.C.: The promotion of self-regulation as a means of enhancing school readiness and early achievement in children at risk for school failure. Child Dev. Perspect. **6**, 122–128 (2012). https://doi.org/10.1111/j.1750-8606.2011.00209.x
27. West, S.G., Finch, J.F., Curran, P.J.: Structural equation models with non normal variables: Problems and remedies. (1995)
28. Chelsea Goforth: Using and Interpreting Cronbach's Alpha. (2015) https://data.library.virginia.edu/using-and-interpreting-cronbachs-alpha/
29. Xie, D., Cobb, C.L.: Revised NEO Personality Inventory (NEO-PI-R). Wiley Encycl. Pers. Individ. Differ. Meas. Assess. 335–350 (2020)
30. Xiong, P., Zhang, J., Wang, X., Wu, T.L., Hall, B.J.: Effects of a mixed media education intervention program on increasing knowledge, attitude, and compliance with standard precautions among nursing students: a randomized controlled trial. Am. J. Infect. Control **45**(4), 389–395 (2017)